Sabrina Schlie

Selective cell control for biomedical applications

Sabrina Schlie

Selective cell control for biomedical applications

Impact of laser-fabricated 3D scaffolds and surface topographies

Südwestdeutscher Verlag für Hochschulschriften

Impressum/Imprint (nur für Deutschland/ only for Germany)
Bibliografische Information der Deutschen Nationalbibliothek: Die Deutsche Nationalbibliothek verzeichnet diese Publikation in der Deutschen Nationalbibliografie; detaillierte bibliografische Daten sind im Internet über http://dnb.d-nb.de abrufbar.

Alle in diesem Buch genannten Marken und Produktnamen unterliegen warenzeichen-, marken- oder patentrechtlichem Schutz bzw. sind Warenzeichen oder eingetragene Warenzeichen der jeweiligen Inhaber. Die Wiedergabe von Marken, Produktnamen, Gebrauchsnamen, Handelsnamen, Warenbezeichnungen u.s.w. in diesem Werk berechtigt auch ohne besondere Kennzeichnung nicht zu der Annahme, dass solche Namen im Sinne der Warenzeichen- und Markenschutzgesetzgebung als frei zu betrachten wären und daher von jedermann benutzt werden dürften.

Verlag: Südwestdeutscher Verlag für Hochschulschriften GmbH & Co. KG
Dudweiler Landstr. 99, 66123 Saarbrücken, Deutschland
Telefon +49 681 37 20 271-1, Telefax +49 681 37 20 271-0
Email: info@svh-verlag.de
Zugl.: Hannover, Leibniz Universität, Dissertation, 2009

Herstellung in Deutschland:
Schaltungsdienst Lange o.H.G., Berlin
Books on Demand GmbH, Norderstedt
Reha GmbH, Saarbrücken
Amazon Distribution GmbH, Leipzig
ISBN: 978-3-8381-1814-7

Imprint (only for USA, GB)
Bibliographic information published by the Deutsche Nationalbibliothek: The Deutsche Nationalbibliothek lists this publication in the Deutsche Nationalbibliografie; detailed bibliographic data are available in the Internet at http://dnb.d-nb.de.

Any brand names and product names mentioned in this book are subject to trademark, brand or patent protection and are trademarks or registered trademarks of their respective holders. The use of brand names, product names, common names, trade names, product descriptions etc. even without a particular marking in this works is in no way to be construed to mean that such names may be regarded as unrestricted in respect of trademark and brand protection legislation and could thus be used by anyone.

Publisher: Südwestdeutscher Verlag für Hochschulschriften GmbH & Co. KG
Dudweiler Landstr. 99, 66123 Saarbrücken, Germany
Phone +49 681 37 20 271-1, Fax +49 681 37 20 271-0
Email: info@svh-verlag.de

Printed in the U.S.A.
Printed in the U.K. by (see last page)
ISBN: 978-3-8381-1814-7

Copyright © 2010 by the author and Südwestdeutscher Verlag für Hochschulschriften GmbH & Co. KG and licensors
All rights reserved. Saarbrücken 2010

1 Outline

1	**OUTLINE**	**1**
2	**ABSTRACT**	**5**
3	**INTRODUCTION**	**7**
3.1	TISSUE ENGINEERING	7
	3.1.1 Scaffold fabrication	9
	3.1.2 Scaffold coating with cells	10
3.2	MATERIAL FUNCTIONALIZATION	11
	3.2.1 Fabrication of defined surface topographies	12
3.3	BIOMATERIAL CELL INTERACTION	13
	3.3.1 Biocompatibility	13
	3.3.2 Adhesion	13
	3.3.3 Adhesion effects on the cytoskeleton	20
	3.3.4 Adhesion correlates with direct gap junction coupling	21
3.4	AIM OF THIS STUDY	23
4	**MATERIALS AND METHODS**	**25**
4.1	LASER TECHNOLOGIES	25
4.2	INVESTIGATED MATERIALS	25
	4.2.1 Polymers and polymer processing	25
	4.2.2 Hydrogels	27
	4.2.3 Metals	28
	4.2.4 Surface coating with adhesion ligands	29
	4.2.5 Material characterization	29
4.3	MATERIALS FOR CELL CULTURE	30
	4.3.1 Sterilization	30
	4.3.2 Cell culture on three-dimensional scaffolds	30
4.4	CELL CULTURE EXPERIMENTS	31
	4.4.1 Cell culture	31
	4.4.2 Analysis of DNA damage effects	31
	4.4.3 Adhesion kinetic	33
	4.4.4 Microscopic analysis	33
	4.4.5 Proliferation assay	36
	4.4.6 Analysis of gap junction coupling	36
4.5	STATISTICAL ANALYSIS	38
5	**RESULTS**	**39**
5.1	CELL RESPONSES TO UNSTRUCTURED MATERIALS	39
	5.1.1 Materials influenced DNA strand breaking	39
	5.1.2 Materials affected adhesion time in a cell specific manner	40
	5.1.3 Materials influenced proliferation in a cell specific manner	40
5.2	CELL RESPONSES TO THREE-DIMENSIONAL SCAFFOLDS	44
	5.2.1 Scaffolds composed of Ormocomp® and PEG SR610	44
	5.2.2 Microscopic analysis of different cell types on three-dimensional scaffolds	45
5.3	CELL TRANSPORT WITH LASER-INDUCED FORWARD TRANSFER	48
	5.3.1 Arrangement of cells in defined pattern	48

        5.3.2   Analysis of DNA damage effects after laser-induced forward transfer 49
        5.3.3   Cell proliferation after laser-induced forward transfer 49
    5.4 CELL RESPONSES TO LASER-FABRICATED SURFACE TOPOGRAPHIES 50
        5.4.1   Surface topographies for material functionalization 50
        5.4.2   Topography induced wettability changes 53
        5.4.3   Topography influenced DNA strand breaking 54
        5.4.4   Topograhical effects on orientation and cell morphology 55
        5.4.5   Topography affected proliferation in a cell specific manner 58
    5.5 ANALYSIS OF ADHESION KINETIC AND PATTERN 60
        5.5.1   Cell specific adhesion kinetic 60
        5.5.2   Cell specific adhesion pattern 61
    5.6 CELL RESPONSES TO ADHESION LIGANDS 62
        5.6.1   Shortterm effects of adhesion ligands 62
        5.6.2   Longterm effects of adhesion ligands 70

6   DISCUSSION 86
    6.1 CELL RESPONSES TO THREE-DIMENSIONAL SCAFFOLDS 86
        6.1.1   Ormocomp® does not negatively affect cellular behavior 87
        6.1.2   The biomedical use of PEG depends on its composition and cell type 87
        6.1.3   Cell localization on three-dimensional scaffolds 89
        6.1.4   Cells adhere on lateral surfaces 90
    6.2 CELL TRANSPORT WITH LASER-INDUCED FORWARD TRANSFER 90
    6.3 SELECTIVE CONTROL OF CELLULAR BEHAVIOR IN DEPENDENCE OF MATERIAL
        CHEMISTRY 91
        6.3.1   Selective control of cellular behavior in dependence of material
                 hydrophobicity 91
        6.3.2   Cell control in dependence of material crosslinking density 93
    6.4 SELECTIVE CELL CONTROL BY SURFACE TOPOGRAPHIES 94
        6.4.1   Selective control of orientation by groove structures 95
        6.4.2   Selective control of cellular behavior by micro-, hierarchical nano- and
                 micro- superimposed-, and nanostructures 96
    6.5 CELL SPECIFICITY OF ADHESION MECHANISM 98
        6.5.1   Adhesion time and pattern are cell specific 99
        6.5.2   Cell specific ligand priority ranking in dependence of the ligand
                 concentration 101
    6.6 THE SELECTIVE CELL CONTROL OF BIOMATERIALS WAS CAUSED BY CELL SPECIFIC
        DIFFERENCES IN ADHESION MECHANISM 105
        6.6.1   Correlation between hydrophobicity and adhesion 105
        6.6.2   Correlation between topography and adhesion 107
    6.7 CONCLUSIONS 108
    6.8 FUTURE PERSPECTIVE 109

7   ATTACHMENT 110
    7.1 REFERENCES 110
    7.2 FIGURES 120
    7.3 TABLES 123
    7.4 ABBREVATIONS 124
    7.5 SOFTWARE 124
    7.6 MEDIA AND SOLUTIONS 125
        7.6.1   Ligand coating 125

	7.6.2	Cell culture	125
	7.6.3	Analysis of DNA damage effects	126
	7.6.4	Staining solutions	126
	7.6.5	Gap junction coupling	126 |//

7.7 ANALYSIS OF CELL MORPHOLOGY – CELL SIZES [μM] 127
7.8 ACKNOWLEDGEMENTS 132

Outline

2 Abstract

In the field of biomedicine and tissue engineering the design and selection of biomaterials are a challenging process. Several criteria have to be taken into account which enable the intended interaction between the material, tissue and cells in dependence of the application. One approach refers to the creation of functioning three-dimensional tissues and cell-coated scaffolds that shall be transplanted to guide tissue formation and reconstruction. In other fields material functionalization is under development. All functionalization methods shall provide a selective cell control followed by an improvement of implant adaptation into the tissue.

This study was focused on the analysis of cell specific responses to different materials and compositions, three-dimensional scaffolds and topographically functionalized materials. Cellular behavior was characterized via DNA damage effects, adhesion, morphology, proliferation, orientation, and gap junction coupling with various cell types. All structures were produced by established techniques at the Nanotechnology Department of the Laser Zentrum Hannover e. V. (Germany) and placed at the disposal for cell experiments performed in this work.

For the fabrication of three-dimensional scaffolds the two-photon polymerization technique was used. Its possibility of high localization of material-light interaction enables the design of any desired three-dimensional microstructure in photosensitive materials with a resolution of 100 nm. To improve the functionality of these scaffolds, there is a demand to coat the entire surface with cells. In dependence of size and structure dimensions cells either fell within the features, lay on the top or adhered on lateral surfaces. However, to overcome a heterogeneous cell distribution on the scaffold, recent advances in biomedical engineering have developed a concept laser-printing such as laser-induced forward transfer. Important preconditions such as the controlled cell arrangement on the target and inhibition of negative transport effects on cellular behavior have to be considered. It was demonstrated that cells could be transported and arranged in defined patterns. Furthermore, this procedure did not harm the cells with respect to DNA strand breaks and proliferation. Therefore, the two-photon polymerization and laser-induced forward transfer are potential technologies for applications in tissue engineering.

The aim of functionalization methods is the imitation of the natural environment of the cells and by this means, the selective control of cellular behavior. With respect to chemical functionalization, the possible use of material hydrophobicity and crosslinking density was investigated. It was shown that cellular behavior can be controlled by both parameters. The second approach referred to a topographical functionalization. Surface structuring was accomplished by femtosecond laser material processing, which enables a flexible and controlled production of micro- and nanostructures in solid materials. The laser-fabricated surface features can also be reproduced via a negative microreplication technique. All investigated topographies showed their effectiveness for selective cell control.

With respect to the results and the fact that adhesion pattern and kinetic were cell specific, it was supposed that the selective cell control of materials is caused by cell specific differences in adhesion mechanism. For this purpose, the influence of four different adhesion ligands from the extracellular matrix on cellular behavior was investigated. It was found that the cells respond to all used ligands with a cell specific priority ranking. Moreover, cell behavior was dependent on ligand concentration. These findings help to explain the observed results and facilitate material search and functionalization for future biomedical applications.

3 Introduction

In the past regenerative medicine, tissue engineering and biomedical research have gained widespread interest and importance due to the increasing lifetime of the population, health problems and diseases followed by rising health expenditures. Therefore, there is a demand to develop therapies and technologies to restore lost, damaged or aging cells, tissues and organs in the human body, to improve surgeries and the quality of life of the patients. Besides pharmacological strategies, a common approach is the fabrication of prothesis or implants used for orthopedic, dental, vascular, cartilage and auditory applications, which shall support or substitute disordered or lost body functions [1-4]. Advances in tissue repair also by implants necessitate biofunctional materials, that not only give cells structural support, but also interact with cells to promote desired biological functions [5].

The design and selection of biomaterials depend on the intended medical application. Development of new biomaterials is an interdisciplinary effort and requires a collaboration between material scientists, engineers, physicists, chemists, biologists and clinicans. A wide variety of materials, synthetic or natural, such as polymers, hydrogels, metals and ceramics are under exploration [6-10]. In order to serve for longer period without rejection an implant should possess several attributes. Mechanical properties such as hardness, tensile strength, modulus, swelling and elongation decide the type of material to be selected. Furthermore, high corrosion and wear resistance determine the longevity of the material [1, 11]. Material characteristics such as chemistry, surface roughness and topography guide implant adaptation, for instance osseointegration [12]. Also material biocompatibility is one important factor.

Before performing clinical investigations, the possible biomedical use of biomaterials is determined by basic research. For a rational design of biomaterials all variables influencing cell functions and tissue morphogenesis have to be considered.

3.1 Tissue engineering

In the field of tissue engineering there is a demand to produce patient-specific substitutes that may serve as alternatives to medical devices, tissue reconstruction and organ transplantation. Since tissues are complex three-dimensional multi-layered structures, the properties of the tissue-engineered constructs have to create an appropiate three-dimensional environment to promote cell function and tissue regeneration [13]. However, the engineered tissue must not only grow to fill a defect and integrate with the host tissue, but often also grow in concert with the changing needs of the body over the time [14]. The necessity of tissue engineering is illustrated by the ever-widening supply and demand dismatch of organs and tissues for transplantation [15]. Hence common implant materials have a limited lifetime, new materials could be a benefit that stimulate the body's own regenerative mechanism, restoring diseased or damaged tissue to its original state and function. By this means re-implantation at older age of the patient could be avoided, and thereby health costs could be decreased.

Tissue-engineered constructs consist of synthetic three-dimensional scaffolds whose structure should mimic in micro- and nanoscale the tissue to be replaced and regenerated. Design criteria refer to the production of a highly interconnected porous networks with pore sizes large enough for fluid and nutrient exchange, vascularization, cell and tissue ingrowth. Scaffolds can be bioactive ceramics, polymers, glasses or nanoscale composites made of synthetic or natural materials. One promising approach is the use of temporary scaffolds which degrade at the same rate the cells produce their own extracellular matrix, the organic template of tissues. By this means the body will then remodel the scaffold conditions into mature tissue [16].

To improve the functionality of tissue-engineered constructs, research has turned towards the creation of cell-coated implants that mimic native tissues with respect to anatomical geometry, cell placement, and microenvironment of the cells [17]. Thereby, the use of autologous cells reduces the risk of immune rejection. An alternative cell source is embryonic stem cells, which can differentiate in any cell type [16, 18].

Parallel to the development of scaffolds, material-cell interactions have to be analyzed. In particular the focus lies on understanding cell processes which are responsible for the formation of three-dimensional tissue constructs, hence cellular behavior pattern in three-dimensional matrices differs from planar two-dimensional cell culture conditions [19, 20].

3.1.1 Scaffold fabrication

Using conventional approaches for scaffold fabrication, such as freeze-drying, liquid-liquid phase separation, solvent casting, electrospinning etc., it is possible to control pore connectivity and pore size. However, no active control over the internal architecture of such scaffold, for instance the size and the position of each individual pore is provided. As a consequence, it is virtually impossible to produce structures in accordance to a predefined design or series of identical scaffolds. Therefore, other technologies are needed that can fabricate scaffolds in a controlled, cost-effective, and reproducible manner. For this purpose, very promising is the use of the two-photon polymerization technique, which enables the design of any desired three-dimensional structure down to a resolution of 100 nm [21, 22]. Taking its origin from multiphoton microscopy, the technique relies on the ability of high localization of the material-light interaction. Using photosensitive materials, this interaction results in a material solidification only within the focus region of the laser beam. By moving the laser focus through the material, any desired three-dimensional structure can be produced (Figure 1). This technique has been established by Dr. A. Ovsianikov at the Laser Zentrum Hannover e. V. (Germany).

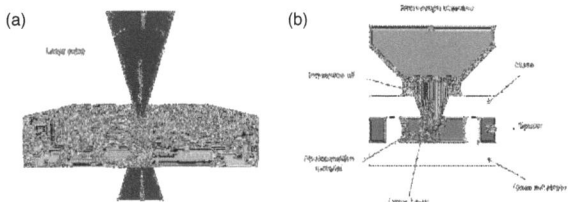

Figure 1: Two-photon polymerization
(a) movement of a laser pulse through a photosensitive material, (b) sample table within the experimental setup.
Images were received from Dr. A. Ovsianikov (Laser Zentrum Hannover e. V., Germany)

Solidification of the material correlates with photocrosslinking of the molecular chains. Photosensitivity can be reached by the application of photointitiators, which generate free radicals by the exposure to UV or visible light in order to initiate the polymerization process [23]. Potential candidates for scaffold fabrication are commercially available photosensitive polymers such as the organic-inorganic hybrid Ormocomp® or epoxy-based SU8 [22] and crosslinkable hydrogels. Especially hydrogels are of interest in the field of biomedicine and tissue engineering with respect to controllable and various chemical and physical properties, the possible combination with bioactive molecules such as growth factors, and material degradation, which enables the design of drug delivery vehicles [11, 15, 23 - 25]. The crosslinking of hydrogels is characterized by the degree of substitution (DS) defined as the average number of substituted hydroxyl groups. This parameter determines the degradation property and material mechanic.

3.1.2 Scaffold coating with cells

Parallel to the development of three-dimensional scaffolds, another challenge relies on seeding cells onto the fabricated constructs, since a sedimentation of the cells has to be avoided. Furthermore, the total scaffold area including lateral surfaces has to be coated with cells. As cell seeding may lead to a heterogeneous cell distribution, recent advances in biomedical engineering have developed a concept of tissue and organ printing [17]. It was shown that different techniques, based on inkjet- and laser-writing technologies, enable the controlled deposition of the support material such as cells to a defined target [26, 27]. One possible method is the laser-induced forward transfer (LIFT), in which a droplet including cells with a forward motive force is ejected from a source substrate and transferred to a target substrate in air [28]. The forward motive force is coming from a shockwave generated in the substrate layer through a local evaporation produced by a focused laser pulse (Figure 2). The laser-induced forward technique has been established by Dipl.-Ing. M. Grüne and Dr. L. Koch at the Laser Zentrum Hannover e. V. (Germany).

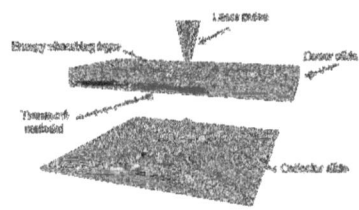

Figure 2: Schematic image of laser printing setup
Image was received from Dipl.-Ing. M. Grüne and Dr. L. Koch (Laser Zentrum Hannover e. V., Germany)

From the technical point of view the challenge of laser-induced forward transfer relies on finding the right laser processing parameters, which enable the controlled transport of the cells to a target and arrange them in a defined pattern, also in three dimensions. From the biological point of the view, this transportation shall not harm the cells with respect to DNA damage effects, proliferation, and other behavior pattern. Besides pre-coating scaffolds with cells, such printing methods could also be applied to create multiple and complex layers of different cell types, for instance useable for skin replacement.

3.2 Material functionalization

The success of an implant is determined by its integration into the tissue surrounding the biomaterial. Not only mechanical and structural properties of the tissue have to be imitated, but also specific cell responses have to be addressed, thus controlling or guiding tissue formation in contact with the biomaterial [29]. The knowledge of cell-biomaterial interactions is a key consideration when developing medical implants and tissue engineering strategies. Especially, the role of fibroblasts is of importance, since these cells participate in foreign body reactions after implantation such as the formation of granulation tissue and fibrosis. Both formations surround the biomaterial at the tissue interface, followed by an implant isolation. In some cases this can reduce implant function and lifetime, and in the worst case, result in re-implantation [30]. For instance, this problem occurs with cochlea implants, which function to restore hearing of deaf patients by electrical stimulation of the auditory nerve is negatively affected by fibroblasts [31]. Research concentrates on the generation of biomaterials, which could control cellular behavior in a cell specific manner – inhibiting fibroblasts while stimulating the competing cell types in dependence of the implant application.

Since conventional biomaterial do not fulfill all specifications with respect to selective cell control, functionalization methods are under development. They implicate changes in material properties such as chemistry and surface topography [32, 33]. Furthermore, a biological approach via material combination with bioactive molecules has been performed [15]. All strategies have in common, that the materials are biologically inspired, and via functionalization copy the natural environment of the cells. In literature, biomaterials that provide a selective control of cell responses are often called 'implants of the next generation' or 'intelligent biomaterials'.

3.2.1 Fabrication of defined surface topographies

Within the tissue, cells interact with micro- and nanoscale topographical projections and depressions that vary in composition, size, and periodicity [34]. Already in the 1950's it was demonstrated that cells react to the topographic structure of their environment [35]. Over the time it could be shown that cell sensitivity to the environment is also reflected by being able to respond to objects as small as 5 nm [36]. Several studies reported that cell behavior can be influenced by surface roughness and structures such as pores, grooves and pits in micro- and nanometer dimensions [37 - 39]. Such defined surface features can be produced due to advances in micro- and nanofabrication.

For a precise design of surface topographies, diverse technologies such as polymer demixing, lithography, plasma treatment, laser irradiation, etching, and other methods are used [34, 38, 40]. Laser processing inducing material ablation, provided for topographical functionalization, has various advantages over methods, namely low surface contamination, low mechanical damage, and controllable surface texturing with complicated geometries in micro- and nanometer scale [41]. Ultrashort pulsed laser processing presents additional benefits due to a better resolution, a reduced heat-affected zone, and a larger variety of surface structures applicable to almost all solid materials [42]. Size dimensions such as height and distance of the generated structures can be varied, controlled, and reproduced by the right laser processing parameters [42]. Additionaly, the negative replication process enables the transfer of the fabricated features into soft materials [43]. Surface structuring by femtosecond lasers and the negative replication technique have been established by Dipl.-Phys. E. Fadeeva at the Laser Zentrum Hannover e. V. (Germany).

Surface characteristics are analyzed via diverse imaging techniques such as scanning electron microscopy and atomic force microscopy, spectroscopy, surface free energy, and wettability [44]. Topographical influences on the wetting properties is of interest, since it determines the surface area for contact. The wetting itself is described by a static contact angle of a water droplet placed onto the surface. Two well-established models by Wenzel [45] and Cassie and Baxter [46] predict that structuring either results in a complete wetting correlating with a decreased contact angle or in an incomplete wetting correlating with an increased contact angle in comparison to the unstructured control surface. An increase or decrease of surface area for contact depends on the provided structure features. The analysis of material wettability in dependence of surface structures has been performed by Dipl.-Phys. E. Fadeeva at the Laser Zentrum Hannover e. V. (Germany).

3.3 Biomaterial cell interaction

The effects of the implant, tissue-engineered constructs and biomaterials in dependence of their properties such as chemistry and topography on the tissue can be differentiated into biological responses characterized by biocompatibility and cellular responses. Last is reflected by influences on cellular behavior estimated by analysis of DNA damage effects, adhesion, orientation, morphology, proliferation, and intercellular communication.

3.3.1 Biocompatibility

Materials used for biomedical applications are expected to be non toxic and should not cause any inflammatory or allergic reactions in the human body. The success of the biomaterial mainly depends on the reaction of the body to the implant and is characterized by biocompatibility. The two main factors that influence the biocompatibility are the host response induced by the material and material degradation in the body environment [1]. According to Anderson [30], biological responses can be separated into the following effects: injury while implantation, blood-material interactions, temporal, accute and chronic inflammation, granulation tissue, and foreign body reactions. In vivo evaluations of tissue responses to the materials such as sensitization, irritation, toxicity, genotoxicity, immune response, and others are important for performance, safety, and regulatory reasons.

3.3.2 Adhesion

The interaction between biomaterials and cells reveal that cellular effects occur in a specific order. As cellular adhesion to the surface is considered to be the first step, the knowledge about cell specificity and material influence on adhesion mechanism may be the key factor to generate a perfectly tissue-integrated biomaterial which controls and guides tissue formation and regeneration [47 - 49]. Adhesion is a very dynamic and complex mechanism mediated by many different factors. First, components of the extracellular matrix associate with the biomaterial surface in a nonspecific manner, whereas the alignment, localization, concentration, and conformation of the components are governed by material properties [50]. Second, several components of the extracellular matrix serve as adhesion ligands, which specifically bind to adhesion receptors localized within the cell membrane. The binding activates intracellular signaling pathways which stimulate cell responses [51, 52]. The whole mechanism regulates cell survival and cell death called anoikis, which is induced upon disruption of the matrix adhesion [53, 54]. Abnormalities in adhesion interaction are often associated with pathological states, including blood clotting and wound healing defects as well as malignant of tumor formation [55, 56]. Because of these significant and wide-ranging regulatory roles, modulating adhesion by biomaterials may provide powerful targets for regenerative medicine and tissue engineering.

Extracellular matrix

The attachment of cells to the extracellular matrix plays a crucial role in the organization, integrity, morphogenesis, and architecture of tissues [57]. Generally, it consists of a complex mixture including glycosaminoglycans, proteoglycans, and proteins. Due to its diverse combination it can either form the interstitial matrix or the basement membrane to anchorage cells, segregate tissues from each other, and regulate intercellular communication [58]. Cell-matrix interaction are mediated by adhesion ligands such as laminins, fibronectin, collagen, vitronectin, and others (Figure 3). By binding to integrins, the primary familiy of adhesion receptors, the attachment initiates signaling cascades involved in the organization of the cytoskeleton, proliferation, migration, and differentiation [59].

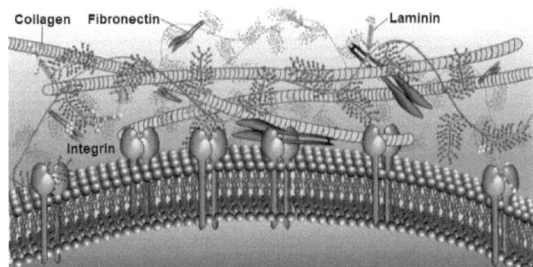

Figure 3: Extracellular matrix including adhesion ligands and receptors
http://219.221.200.61/ywwy/zbsw(E)/edetail4.htm

Introduction

According to Tzu [60] laminins are extracellular heterotrimeric glycoproteins of 400-900 kDa composed of various combinations of α, β, and γ chains. So far, five α, four β, three γ, and totally 16 known laminins have been identified and numbered in the order in which they were discovered. Each chain consists of rodlike, globular, and coiled regions held together by disulfide bonds. Furthermore, laminin molecules can undergo multiple post-translational modifications. The largest chain is the α chain, which contains a long arm at the C-terminus involved in the interaction with the adhesion receptors integrins. The N-terminus can be diverse in length and binds to other laminins to produce calcium-dependently a supramolecular network. At least nine integrins have been described to bind laminin [57, 60 - 65].

Fibronectin is a multifunctional elongated and flexible glycoprotein which exists as a soluble plasma protein and as a fibrillar component of the extracellular matrix. It consists of two similar or identic subunits of 220 kDa held together by disulfide-bonding near the carboxyl terminus. Each of these subunits includes several distinct functional domains which in turn contain three types of structural modules referred as type I, II and III [66]. Former known is a collagen-binding domain of 40 kDa with an alternative affinity to bind gelatin near the amino-terminus, but with no function in mediating cell interaction. Furthermore, multiple heparin-binding reagions near the carboxyl terminus were described which play important roles in the structural organization of the extracellular matrix [67]. The cell-binding reagion of fibronectin with 15 kDa is placed within the type III module. There, the strong and noncovalent binding to adhesion receptors of the cells is mediated by the integrin recognition motif called RGD-sequence (the tripeptide Arg - Gly - Asp), presented as a loop binding to a shallow crevice located between the integrin subunits [68]. Eight integrins have been found to bind fibronectin [62, 66, 69 - 73].

Collagens are large, triple-helical proteins that form fibrils and network-like structures. The helix consists of α-chains with a primary GXY structure (Gly - Xaa -Yaa) repeated several times. Thereby, glycine is placed in the central part of the helix. X and Y are often represented by proline residues, also posttranslational hydroxylated. According to Heino [74] collagens have been numbered from I to XXIX and divided into several subfamilies. Fibril-forming collagen such as I-III, V, XI, XXIV and XXVII have a long and continuous helical domain and are responsible for the tensile strength of the tissue found in bones and cartilage. Fibril-associated collagens like IX with interruptions in their triple helix mediate interaction of fibrils with other macromolecules from the extracellular matrix. Other subgroups function as beaded filaments (IV), anchoring fibrils (VII), networks (IV, VIII, X), and structural proteins in basement membranes (XV, XVIII). Also integral membrane proteins have been described (XIII, XVII, XXIII, XXV). Several different receptors can bind collagen. Concerning integrins, the binding motif is typically a GXX'GER or a DGEA (Asp - Gly - Glu - Ala) sequence recognized by at least five different receptors [61, 62, 69, 71, 75, 76].

Vitronectin, a 70 kDa adhesive glycoprotein, can be found in the extracellular matrix and in plasma at concentrations of 200 - 400 µg/ml. Similarly to fibronectin, it is composed of several functional domains including the somatomedin B domain near the N-terminus, the tripeptide RGD sequence (Arg - Gly - Asp) which binds to integrins and two hemopexin-like domains [77]. It interacts with various proteins to regulate numerous cell functions. In dependence of the bound protein, the effects of vitronectin involve calcium signals [78]. The binding to at least four different integrins promotes cell adhesion to the extracellular matrix [69, 71, 79].

Adhesion receptors

The described proteins laminin, fibronectin, collagen, and vitronectin serve as ligands to adhesion receptors called integrins. This name denotes the integral membrane nature of the transmembrane receptor and its role in the integrity of the extracellular matrix to the cytoskeleton via integrin clustering to the formation of focal adhesions [51, 80]. Functional integrin receptors are dimers of α and β subunits. So far, 18 α and 8 β subunits forming 24 different integrin dimers have been identified [56]. According to Siebers [71] the length of the α chain is 1008-1152 aminoacids and the β chain around 770, with a cytoplasmatic reagion of 22 - 29 and 20 - 50, and a transmembranous part of 20 - 29 and 26 - 29, respectively.

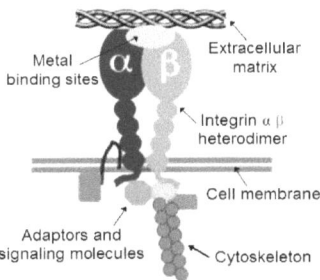

Figure 4: Schematic image of integrin receptors
http://dan1.medkem.gu.se/program_files/image004.gif

As shown in Figure 4, in the extracellular part integrins present a ligand binding 'head', whereas the ligand-binding region is located in an inserted (I) domain, inserted in a G protein-like seven-bladed β-propeller domain, within the α chain or within a structural contribution of both α and β chain [81]. The I-domain can present two different conformations: open (high affinity) and closed (low affinity), regulated by divalent cations placed in the MIDAS region (metal ion dependent adhesion site). Furthermore, the I-domain of β chain contains beside the MIDAS also the ADMIDAS (adjacent to MIDAS) and the LIMBS region (ligand induced metal binding site). It was shown that Mn^{2+} and Mg^{2+} promote ligand binding to RGD sequences while Ca^{2+} has an inhibitory effect [70, 81 - 83]. Integrin conformation and activation are dependent on the β chain. So far, little is known about the transmembrane domains. The cytoplasmic parts of α and β chains are α-helical and miss enzymatic features. Some data support that a close association of α and β chains keeps the integrins in a resting state.

Concerning possible ligands that bind to integrins, there is an overlap in specificity, with many integrins capable of binding to more than one protein, whereas proteins can act as ligands for more than one integrin. After the binding of the ligand, which anchorages the cells in the extracellular matrix, integrins cluster into focal contacts, consisting of additional cytoskeletal proteins, adaptor molecules, and kinases, followed by the activation of diverse signaling cascades. Integrin communication over the plasma membrane in both directions, to the extracellular matrix and to the intracellular part, can be distinguished between outside-in and inside-out signaling [81].

Adhesion signaling

The inside-out signaling orginates from non-integrin surface receptors or cytoplasmic molecules that activate and deactivate integrins. Besides the prooved regulatory role of divalent cations in the extracellular region of integrins, Gumbiner [58] and Gahmberg [81] suggested talin, α-actinin, paxillin, filamin, integrin-linked kinase (ILK), and focal adhesion Tyr kinase also to be involved caused by a direct influence on the cytoplasmatic part of the β chain. The activation of integrins correlates with the binding affinity to adhesion ligands and depends on integrin clustering and conformational changes in the integrin structure [84].

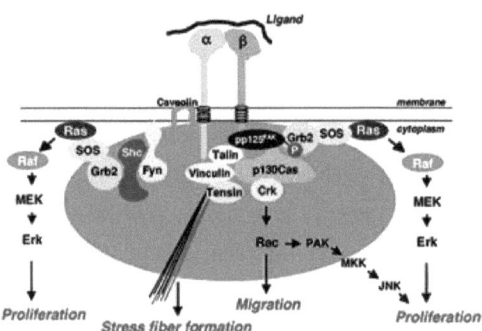

Figure 5: Schematic image of outside-in signaling in adhesion mechanism
www.charite.de/.../images/danker_schema1.gif

A complex series of steps leads from the initial integrin interaction with ligands from the extracellular matrix to transmembrane effects stimulating diverse signaling pathways shown in Figure 5 which activate the organization of the cytoskeleton, proliferation, migration, differentiation, and gene expression [59]. This signaling machinery crucial for cellular behavior and responses to the substrate is called outside-in signaling and is caused by the formation of focal contact complexes.

After the connection to the extracellular matrix, large molecular complexes link to the integrins at the cytoplasmatic side to form focal adhesions, which are dynamic and heterogeneous structures. So far, more than 50 focal adhesion molecules have been identified. In contrast to the extracellular and transmembrane regions consisting of adhesion ligands and integrins, the cytoplasmatic region of focal adhesions is very diverse and complex. Basically, it can be divided into three functional groups. First, into structural molecules from the cytoskeleton such as actin, talin, tensin, vinculin, and others. Second, into enzymatic molecules like protein tyrosine kinases namely focal adhesion kinase (FAK), Src, Fyn, and others, like protein serin/threonine kinases namely integrin-linked kinase (ILK), and others, like protein phophatases, and modulators of small GTPases. Third, several adaptor molecules are involved such as paxillin, Grb, Crk, Cdc 42, and Shc [85].

The association of cytoskeletal molecules like talin, vinculin, and tensin into focal adhesions serves as a positive feedback system, as actin filaments are reorganized into stress fibers which promote integrin clustering and enhance extracellular matrix binding [86].

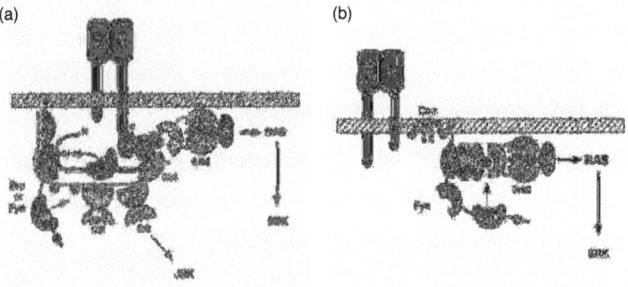

Figure 6: Model of (a) FAK and (b) Shc pathways [88]

Clustering focal adhesion kinase (FAK) into focal adhesions enhances this autophosphorylation at Tyr^{397} creating a binding side for the SH2-domain receptor protein Src followed by further phosphorylations at Tyr^{925}. The role of Src seems contradictory, as it not only initiates binding sites at FAK for Grb2 activating Ras and mitogen-activated protein kinase (MAPK) signaling cascades, but also excludes FAK from the focal complex under certain conditions [87]. Furthermore, Src stabilzes paxillin and tensin needed for cell spreading and migration by recruiting the adaptor protein Crk [88]. The combination of FAK with Grb2 and the exchange factor SOS are former known to modify the cytoskeleton needed for its dynamic features [51]. Additionally, the signaling pathways over FAK and Src lead to c-Jun amino-terminal kinase (JNK) and Ras stimulations followed by activations of extracellular signal-regulated kinase (ERK) and MAPK involved in cell cycle progression [52]. Clark [51] also mentioned a Ras-MAPK cascade inducing cytosolic phospholipase A_2 activation which liberates arachidonic acids and its metabolites from glycerolphospholipids. The connection of phosphoinoside 3-kinase (PI 3K) to FAK phosphorylates Akt which thereby inactivates two pro-apoptotic proteins such as Bad and caspase-9 [86]. The loss of attachment to the extracellular matrix causes an imbalance of PI 3K and Akt resulting in apoptosis called ankoikis [89]. (Figure 6 a)

In addition to activating FAK, integrins affect tyrosine kinase Fyn which phophorylates the adaptor protein Shc caveolin-1 dependently. This complex can be linked to Grb2-SOS. As soon as Shc is connected to the focal adhesion, it leads to the activation of Ras-extracellular signal-regulated kinase (ERK) and mitogen-activated protein kinase (MAPK), which is proportional to the binding density to the extracellular matrix [90]. Even though both kinases are also stimulated by FAK, Shc activation appears to play a more important role [86]. The induced functions of ERK and MAPK are very divers and complex. Tzu [60] demonstrated their influence on migration. More pronounced is their control of cell cycle progression. (Figure 6 b)

The activated ERK regulates the signal transduction at the major cell cycle checkpoints such as G1/S, G2/M, and M/G1 followed by a stimulation of transcription, translation, and cell cycle progression [56, 86, 89, 91].

3.3.3 Adhesion effects on the cytoskeleton

Adhesion is the result of clusters of membrane-spanning integrin receptors that link the extracellular matrix to cytoskeletal elements. Several fundamental features which determine tissue function and integrity are related to the cytoskeleton such as cell shape control, cellular mechanic, cell mechanochemistry, cell volume regulation, migration, cell spreading, apoptosis, and others [92 - 95]. Basically, the cytoskeleton is composed of microtubules, interconnected microfilaments, and intermediate filaments which undergo continuous and dynamic changes in their structure. These changes are associated with the formation, organization, and remodelling of matrix contacts.

When cells come in contact with the extracellular matrix, different morphological characteristics and contact types occur. Filopodia are considered to be the first type, since they have primary functions in sensory guidance, adhesive selection, and the integrin-rich composition enables the binding to the surface and formation of focal adhesion complexes [93]. They have an average diameter of 0.2 - 0.5 μm and are 20 - 200 μm long consisting of actin filaments. According to Gahmberg [81] Cdc 42 activated via integrin-FAK association is responsible for filopodia formation. The initial matrix attachment also includes the formation of spikes which are with 2 - 10 μm a lot shorter than filopodia. A subgroup of spike formations are called hemidesmosomen which require $α_6β_4$ integrin and connect to the intermediate filaments. Their particular function appears to be in cell motitily [93]. Afterwards, lamellipodia are generated and arranged between filopodia. Even though these extensions also consist of actin, their functional role is distinct when compared with filopodia [96]. They are involved in cell spreading and migration. This observation refers to the tensegrity model established by Ingber [94]. Cell type dependently, further podosomes, invadopodia, and pseudopodia can be formed [93]. Stable matrix constructs correlate with the development of focal adhesion complexes associated with cytoskeletal molecules such as talin, vinculin, and tensin which initiate the formation of actin stress fibers to enhance the binding to the extracellular matrix [86].

Except for hemidesmosomen, the molecular basis of all contacts between cells and the extracellular matrix requires actin. With respect to the outside-in signaling of adhesion mechanism, the polymerization of actin is initiated by focal adhesion kinase (FAK) [62]. Clark [51] suggested that the rearrangement of actin induced by integrins is Ras-independent. Several studies revealed that it rather correlates with Rho familiy members such as Cdc 42 and Rac [52, 58, 81].

3.3.4 Adhesion correlates with direct gap junction coupling

The formation of gap junction channels generates a direct contact between the cytoplasma of neighboring cells. Thereby, electric signals, hormons, second messengers, and metabolites are exchanged from cell to cell to promote physiological activities which are essential for the formation of real functioning tissues. Gap junction channels are former known to have a specific role in diseases and modulate cellular behavior such as cell cycle progression, differentiation, and apoptosis [97].

Two half-channels called connexons within the cell membrane associate to a cell-to-cell channel and allow the transition of molecules smaller than 1 kDa [98, 99]. Each connexon is built by six connexins that oligomerize (Figure 7). Connexins constitute a familiy of more than 20 homologous proteins in human that are temporally and spatially distributed throughout the body. They are numbered with suffixes referring to the molecular mass in kDa. Gap junction functions are dependent on the connexin expression and can further be influenced by posttranslational modifications such as phosphorylations of connexins.

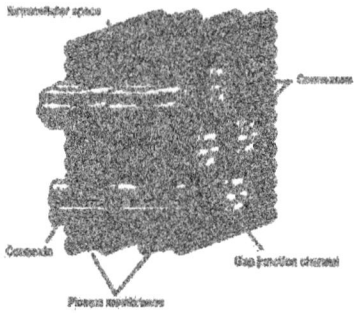

Figure 7: Schematic image of gap junctions
http://img.tfd.com/dorland/thumbs/junction_gap.jpg

Gap junction coupling is also necessary for cell binding to the extracellular matrix, since migration, the formation of focal contacts, and the initiated signaling cascades require second messengers such as Ca^{2+}. The outgrowth of lamellipodia is regulated by Ca^{2+}-signals [83]. Migration as a result of attachment and deattachment of lamellipodia to the extracellular matrix occurs by destabilizing focal complexes and contractile forces modulated by Ca^{2+} [100]. Furthermore, it was shown that migration is facilitated by the presence of connexins [101]. According to Conklin [83] the effect of transient fluxes of Ca^{2+} on focal complexes is significant as integrins have no catalytic activity, which is needed for the autophosphorylation of focal adhesion kinase (FAK). Moreover, the association of Ca^{2+} to the metal binding sites of integrins in the extracellular region is former known to have an inhibitory effect and decreases the ligand-binding affinity [81, 82].

Imbeault [101] revealed that the extracellular matrix alters connexin expression and / or stability, intracellular distribution, hemichannels, and functional channel activity. In dependence of the adhesion ligand and connexin type it was demonstrated that connexin expression can either be up- or down-regulated. For instance, Lampe [102] showed that the upregulation of connexin 43 is Rho familiy-dependent which is activated via focal adhesion complexes. Integrin linked kinase (ILK) which interacts with ERK participates in the down-regulation of connexin 32 [103]. Moreover, the extracellular matrix is involved in posttranslational modifications of connexins [104]. Since connexins are co-localized with actin filaments, a real time control of gap junction coupling by mechanical forces was suggested [103, 105]. These forces are induced by contacts to the extracellular matrix and the formation of focal adhesion complexes [94].

These information point out a clear correlation between adhesion to the extracellular matrix and gap junction coupling.

3.4 Aim of this study

In this work, a wide range of biomaterials and adhesion ligands were investigated with focus on cellular behavior. On the one hand the materials were used to generate three-dimensional scaffolds by two-photon polymerization technique, established by Dr. A. Ovsianikov. On the other hand defined surface topographies were fabricated by femtosecond lasers and negative replication process, established by Dipl.-Phys. E. Fadeeva. Laser processing was performed at the Laserzentrum e. V. in Hannover (Germany).

Cellular responses to the unstructured materials, tissue-engineered constructs, and adhesion ligands were characterized by the parameters DNA damage effects, adhesion, morphology, orientation, proliferation, and gap junction coupling. Since the measurements were performed with different cell types such as GFSHR-17 granulosa cells, human fibroblasts, NIH3T3 fibroblasts, SH-SY5Y neuroblastoma cells, GM-7373 endothelial cells, HaCaT keratinocytes, MG-63 osteoblasts, A10 smooth muscle cells, human and porcine mesenchymal stem cells, cell specific effects of the materials could be evaluated.

First, the influence of untreated materials such as polymers (Ormocomp®, silicone elastomer), hydrogels (hydroxymethacrylathydroxyethylstarch (HESHEMA), poly (ethylene glycol) diacrylate (PEG)), and metals (silicon, platinum, titanium) was tested. Ormocomp® and PEG were prepared by Dr. A. Ovsianikov, silicon and metals by Dipl.-Phys. E. Fadeeva (both Laser Zentrum Hannover e. V., Germany). HESHEMA was received from the Institue of Technical Chemistry (TU Braunschweig, Germany). In dependence of material properties and compositions like wettability, degree of substitution, molecular weight, applied photoinitiator, washing, and aging cell specific responses were estimated.

The two-photon polymerization technique enables the design of three-dimensional scaffolds. It was performed by Dr. A. Ovsianikov (Laser Zentrum Hannover e. V., Germany). Coming to three-dimensional structures the questions were, whether the cells fall within the features, lay on the top or adhere on lateral surfaces, whether they present their normal morphology and are able to proliferate. Furthermore, a correlation between cellular behavior and scaffold dimensions was addressed.

Parallel to the development of tissue-engineered constructs, there is a demand to pre-coat scaffold with cells. Therefore, it was analyzed, whether the laser-induced forward transfer (LIFT) is a possible tool to transport cells, arrange them in defined pattern, and if the transfer itself affects cellular behavior. Laser-induced forward transfer was provided by Dipl.-Ing. M. Grüne and Dr. L. Koch (Laser Zentrum Hannover e. V., Germany).

For material functionalization surface topographies such as micrometer spikes and grooves, hierarchical nano- and micro- superimposed structures, and nanogrooves and -roughness were generated. The structures were fabricated via ablation with femtosecond lasers or with the help of the negative replication technique. Material structuring was performed by Dipl.-Phys. E. Fadeeva (Laser Zentrum Hannover e. V., Germany). Their potential for a cell specific control of cellular behavior was investigated.

To improve implant adaptation, a biomaterial has to provide selective cell control. By this means, it was proposed that a possible control can be caused by cell specific differences in adhesion mechanism which thereby can be influenced selectively by material properties. Disparities of adhesion mechanism were analyzed with focus on adhesion time and pattern, and cell specific responses to adhesion ligands such as laminin, fibronectin, collagen type I, and vitronectin in dependence of their concentrations.

4 Materials and Methods

4.1 Laser technologies

Three different laser technologies were used in this study to generate three-dimensional tissue-engineered constructs, to transfer cells to a defined target applicable for pre-coating scaffolds with cells, and for the fabrication of surface topographies by means of material functionalization. All techniques were performed at the Laser Zentrum Hannover e. V. (Germany) at the Nanotechnology Department under the supervision of Prof. Dr. B. Chichkov. The two-photon polymerization technique enabling the design of scaffolds was carried out by Dr. A. Ovsianikov. Dr. L. Koch and Dipl.-Ing. M. Grüne were responsible for cell transportation via laser-induced forward transfer method. Dipl.-Phys. E. Fadeeva provided surface structuring by femtosecond lasers and negative replication process.

Detailed descriptions of the experimental setups, the laser processing parameters, and the accomplishment of the negative replication technique were reported [22, 23, 43 – 45, 106].

4.2 Investigated Materials

4.2.1 Polymers and polymer processing

Ormocomp® (Organically Modified Ceramics)

The hybrid organic-inorganic polymer Ormocomp®, a member of the Ormocer® family (Microresist Technology GmbH, Germany) includes urethane- and thioether (meth)-acrylate alkoxysilanes, which provide strong covalent bonds between the components. This cross-linking leads to the formation of three-dimensional networks, which can be varied by changing the ratio of organic and inorganic network density. Therefore, it is possible to regulate the desired mechanical, optical, chemical and surface properties.

In this study, the liquid and photosensitive Ormocomp® containing 1.8 % photoinitiator Irgacure 369 (Ciba Specialty Chemicals, Basel, Switzerland) was locally transferred into the solid phase through a free-radical polymerizaiton reaction. For the fabrication of three-dimensional scaffolds, solidification of the material only occurs within the focus region of the laser beam. Flat Ormocomp® surfaces were produced via spin-coating and UV illumination onto glass slides. After irradiation, the non-solidified material was removed by a 1:1 solution of 4-methyl-2-penthanone and 2-propanol. All Ormocomp® samples were produced by Dr. A. Ovsianikov at the Laser Zentrum Hannover e. V. (Germany).

The two-photon polymerizaiton technique was used to produce gratings with different size dimensions and cylinders out of Ormocomp®. A microscopic study was performed with human fibroblasts and SH-SY5Y neuroblastoma cells to oberseve cell localization on the three-dimensional scaffolds. Parallel to the development of scaffolds in micrometer scale, flat samples were used to characterize material effects on cells in general with focus DNA strand breaking and proliferation. GFSHR-17 granulosa cells, SH-SY5Y neuroblastoma cells, and GM-7373 endothelial cells were used.

Silicone elastomer

A two-component silicone elastomer MED-4234 (NuSil Silicone Technology, Cindy Lane Carpinteria, USA) was mixed in proportion of 10:1 according to product description profile. For the experiments three different procedures were used to prepare silicone elastomer samples. First, via spin-coating flat samples were created to investigate biomaterial cell interaction in general. Second, these flat samples were needed to generate surface topographies in micrometer scale such as so-called spike structures with the help of femtosecond lasers. Third, silicone elastomer was poured over other laser fabricated surface features for negative replication process. All silicone elastomer samples were produced by Dipl.-Phys. E. Fadeeva at the Laser Zentrum Hannover e. V. (Germany).

These different types of samples were used to measure proliferation profiles, to characterize material influences on DNA strand breaking, cell morphology, and adhesion of human fibroblasts, GM-7373 endothelial cells, and SH-SY5Y neuroblastoma cells with respect to cellspecific responses.

4.2.2 Hydrogels

Hydroxymethacrylathydroxyethylstarch (HESHEMA)

The hydrogel hydroxymethacrylathydroxyethylstarch (HESHEMA) was synthesized at the Institute of Technical Chemistry (TU Braunschweig, Germany) according to previous descriptions [107]. Shortly, hydroxyethylstarch (HES) solved in DMSO (1:8) was mixed with a hydroxyethylmethacrylate solution (HEMA). Different ratios of HEMA and HES produce crosslinkable HESHEMA derivatives with variable degree of substitutions (DS).

For this work, three different HESHEMA derivatives with variable DS values (0.07, 0.11 and 0.2) were synthesized. Afterwards, HESHEMA was dissolved in distilled water (10 wt%) and stirred at room temperature in the dark for three hours. Then 0.1 wt% photoinitiator Irgacure 2959 (Ciba Speciality Chemicals, Basel, Switzerland) was added. After another hour of stirring, HESHEMA solution was distributed on sterile glass slides and polymerized under UV light for about 30 minutes.

To analyze HESHEMA effects in dependence of the crosslinking property on cellular behavior, adhesion kinetic and proliferation profiles of human fibroblasts, GM-7373 endothelial cells, and SH-SY5Y neuroblastoma cells were examined.

Poly(ethylene glycol) diacrylates (PEG)

Acrylated poly(ethylene) glycols can be used to produce photo-crosslinkable hydrogels. The biomedical application of PEG-based photosensitive materials was studied on the example of two different PEGda materials having molecular weights of 302 (SR259, Sartomer) and 742 (SR610, Sartomer). In order to obtain a photopolymerizable composition, one photoinitiator was added to a final concentration of 2 wt%. The commercially available photoinitiators 4-bis diethyl-aminobenzophenone (Bis, Sigma-Aldrich, Taufkirchen, Germany) and Irgacure 2959 (Irg, Ciba Speciality Chemicals, Basel, Switzerland) were introduced for comparison. All PEG samples were prepared by Dr. A. Ovsianikov at the Laser Zentrum Hannover e. V. (Germany).

First, PEG pellets with a diameter of 6 mm and thickness of 1 mm were prepared by photopolymerization with UV light. On the one hand DNA damage effects of PEG pellets in dependence of the molecular weight (SR259 and SR610) were analyzed with GFSHR-17 granulosa cells. Both samples were supplemented with 2 wt% photoinitiator Bis. On the other hand the influence of fresh and aged PEG pellets (SR610, both supplemented with 2 wt% photoinitiator Irgacure 2959) was determined. Material aging was accomplished by putting fresh samples into destilled water for seven days. Cell responses to PEG SR610 (2 wt% photoinitiator Irgacure 2959) were characterized via DNA damage effects, proliferation, and adhesion kinetic of human fibroblasts, GM-7373 endothelial cells, and SH-SY5Y neuroblastoma cells. Second, photostructuring of PEG was reached by means of two-photon polymerization technique. Scaffolds with different size dimensions such as heigth and diameters were generated. These samples were used for a microscopic study with NIH3T3 fibroblasts and GM-7373 endothelial cells.

4.2.3 Metals

Silicon

Single-crystal p-type silicon (110) samples were used to generate surface structures in micrometer scale. After femtosecond laser irradiation, the samples were treated using a 10 % hydrofluoric acid (HF) aqueous solution to remove oxide layer on the surface. After the washing step with HF, several washing procedures with water followed. The produced structures (so-called spikes) also served as master copies for negative replication process with silicone elastomer. The samples were prepared by Dipl.-Phys. E. Fadeeva at the Laser Zentrum Hannover e. V. (Germany). Surface structure effects on cellular behavior were analyzed by DNA damage effects, morphology, and proliferation. All cell experiments were performed with human fibroblasts and SH-SY5Y neuroblastoma cells to figure out cell specific responses.

Platinum

From a commercial rolled platinum foils with a purity of 99.99 % (Goodfellow, Ltd) platinum samples with a size of 5 x 5 x 0.25 mm were prepared. The first laser-manufactured topography was a periodic surface grating in nanometer scale. The second surface type was a combination of random nano- and micro-roughness. The samples were prepared by A. Y. Vorobyev (University of Rochester, USA). Platinum samples were used to determine DNA damage effects and proliferation of human fibroblasts.

Titanium

Titanium samples with the dimensions of 3 x 3 x 1 mm were applied to generate different surface structures in micrometer scale. Before structuring, the samples were mechanically polished and further cleaned with acetone followed by methanol. In addition to hierarchical nano- and micro- superimposed structures, also periodic gratings with different size dimensions were produced with the help of femtosecond lasers. All samples were prepared by Dipl.-Phys. E. Fadeeva at the Laser Zentrum Hannover e. V. (Germany). The measurements were performed with human fibroblasts and MG-63 osteoblasts to investigate cell specific responses with respect to orientation, proliferation, and DNA damage effects.

4.2.4 Surface coating with adhesion ligands

In order to investigate cellular adhesion mechanism, different adhesion ligands from the extracellular matrix were used. Collagen type I solution from rat tail, laminin from Engelbreth-Holm-Swarm murine sarcoma basement membrane, fribonectin from bovine plasma and vitronectin from bovine plasma were all purchased from Sigma-Aldrich (Taufkirchen, Germany). Following the production description, for each ligand a sterile stocking solution of 0.01 % in phosphate buffer salt (PBS) was prepared. Only collagen was solved in sterile distilled water. The stocking solutions were stored at - 20 °C. One day before use sterile glass slides were coated with the stocking solutions at different concentrations. For collagen concentrations of 10 μg/cm², 8 μg/cm², and 6 μg/cm², for laminin 2 μg/cm² and 1 μg/cm², for fibronectin 5 μg/cm², 3 μg/cm², and 1 μg/cm² were prepared. According to the description profile vitroncetin could only be used at 0.1 μg/cm². The slides were kept at room temperature over night and rinsed with PBS before starting the measurements. Ligands effects on human fibroblasts, GM-7373 endothelial cells, SH-SY5Y neuroblastoma cells, HaCaT keratinocytes, MG-63 osteoblasts, and A10 smooth muscle cells were documented by adhesion kinetic, adhesion pattern, morphology, proliferation, and gap junction coupling. The shortterm measurements with the maximum ligand concentrations were restricted to serum-free cell culture media over a cultivation time of five hours. The longterm measurements were performed with serum-containing cell culture media in dependence of the ligand concentration.

4.2.5 Material characterization

The investigated materials were received from the Laser Zentrum Hannover e. V. (Germany) and the Institue of Technical Chemistry (TU Braunschweig, Germany), which also analyzed material chemistry. Furthermore, a correlation between surface structuring and wettability was addressed, which was performed by Dipl.-Phys. E. Fadeeva at the Laser Zentrum Hannover e. V. (Germany). A detailed description of the diverse experimental procedures have been published [23, 41, 42, 106].

4.3 Materials for cell culture

4.3.1 Sterilization

All the materials, scaffolds, surface structures, and adhesion ligand-coated substrates were sterilized under UV light for at least 30 minutes.

4.3.2 Cell culture on three-dimensional scaffolds

Parallel to the development of scaffolds, an understanding of how cells interact with three-dimensional features is of great interest. One question is whether the cells are able to adhere on lateral surfaces.

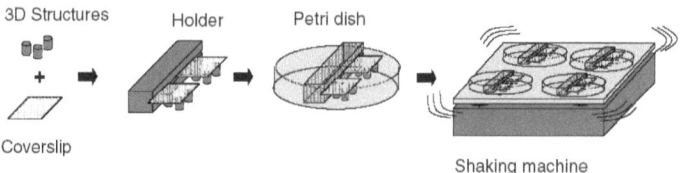

Figure 8: Shaking table for three-dimensional structures

Cylindrical structures composed of Ormocomp® were generated via two-photon polymerization technique directly onto glass slides by Dr. A. Ovsianikov at the Laser Zentrum Hannover e. V. (Germany). These slides were inserted into a plexiglas holder with a structure orientation upside down. Everything was placed inside a petri dish with a diameter of 60 mm containing 10 ml cell culture medium. Afterwards SH-SY5Y neuroblastoma cells were seeded out. To avoide the sedimentation of the cells, the petri dishes were placed on a shaking table within the cell incubator (Heraeus, Hanau, Germany) to keep the cells in suspension (Figure 8). For observation a Nikon stereo microscope (Nikon, Düsseldorf, Germany) was used.

4.4 Cell culture experiments

4.4.1 Cell culture

Material effects on cellular behavior were studied using GFSHR-17 granulosa cells, human fibroblasts, NIH3T3 fibroblasts, SH-SY5Y neuroblastoma cells, GM-7373 endothelial cells, HaCaT keratinocytes, MG-63 osteoblasts, A10 smooth muscle cells, human and porcine mesenchymal stem cells. The cells were cultivated on the samples or on control glass slides (Thermo Scientific, Karlsruhe, Germany) either in petri dishes with a diameter of 35 mm or in 24-well plates (both from Sarstedt, Nümbrecht, Germany) filled with 2 ml of Dulbecco's modified medium (DMEM; Sigma-Aldrich, Taufkirchen, Germany) supplemented with antibiotics (pH 7.4; 300 ± 5 mosmol). The concentration of fetal calf serum (FCS) was adjusted to the cell type and experiment. While granulosa cells, endothelial cells, and keratinocytes were cultivated in 5 % FCS, for the other cell types a final concentration of 10 % was applied. The shortterm adhesion measurements were performed in serum-free media. The dishes and multi-well plates were placed in a cell culture incubator (Heraeus, Hanau, Germany), in which a 95 % : 5 % air : CO_2 atmosphere, 37 °C and 80 % humidity were maintained. The culture medium was renewed every 2-3 days.

As soon as a monolayer was formed, the culture media was removed and replaced with a 0.25 % trypsin solution solved in PBS (pH 7.4) to detach the adherent cells from the culture surface. After several minutes incubation time at room temperature, fresh culture media was added. The cell suspension was collected and centrifuged at 800 g for 10 min. Then the supernandant was removed and the pellet was resuspended in culture media. To determine the cell density of the suspension, a Fuchs Rosenthal cell counter device or a cell counter Casy TT® (Innovatis, Bielefeld, Germany) were used. After that the cell suspension was used to start a new passage or to begin the experiments.

4.4.2 Analysis of DNA damage effects

In order to determine, whether the materials affected the DNA of the cells, DNA strand breaking of different cell types grown on the samples and under control conditions was analyzed using the comet assay. Comet assay experiments were performed according to previous description [22]. After a cultivation time of 24 h, the cells were trypsinzed, collected and centrifuged at 800 g for 10 min. The pellets were resolved in PBS to

Materials and Methods

2×10^6 cells/ml. Later 50 µl of the cell suspension was mixed with 100 µl of low melting agarose (0.6 %). A 100 µl of this mixture was given onto agarose-coated glass slides and covered with a cover slip. The slides were conserved for solidification at 4 °C for 10 min. Then the cover slip was removed and further 100 µl of agarose was added. After solidification at 4 °C, the slides were incubated in a lysis buffer for 90 min, containing 2.5 M NaCl; 100 mM Na_2EDTA; 10 mM Tris; 1 % lauryl sarcosin; 1 % Triton X-100; 10 % DMSO; pH 10. Subsequently, the cover slips were placed in a horizontal gel electrophoresis chamber, filled with electrophoresis buffer for alkaline comet assay (1 mM Na_2EDTA; 300 mM NaOH; pH > 13). After 40 min adaptation to the buffer, electro-phoresis was performed (25 V; 300 mA; 4 °C; 20 min). For neutralization, the slides were washed three times with Tris-buffer (400 mM Tris; pH 7.4) and dried at room temperature. Comets were visualized by ethidium bromide staining (20 µg/ml) and examined with a fluorescence microscope (Zeiss, Oberkochen Germany), using a xenon lamp and ethidium bromide filter set (excitation at $\lambda = 520$ nm). The images were recorded with a CCD Camera ('Xaw TV'). For a quantitative analysis of the DNA damages such as single and double strand breaks, the tailmoment was used. This parameter is defined as the amount of DNA damages, which can be evaluated by comet scoring software (http://www.autocomet.com/home.php). The results were given as mean of tailmoment ± SEM (n = 4). At least 1000 cells per treatment were evaluated. Comet and software images, which also include other parameters for quantification, are shown in Figure 9.

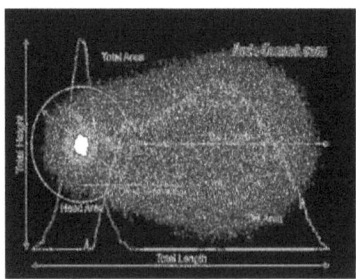

Figure 9: Comet assay parameters according to *autocomet.com*

Comet assay measurements were performed with different cell types such as GFSHR-17 granulosa cells, human fibroblasts, SH-SY5Y neuroblastoma cells, NIH3T3 fibroblasts, HaCaT keratinocytes, human and porcine mesenchymal stem cells cultivated on Ormocomp®, PEG, silicone elastomer, silicon, titanium, and platinum in dependence of material chemistry and topography, and after the laser-induced forward transfer method.

4.4.3 Adhesion kinetic

For a biophysical characterization of adhesion mechanism, adhesion kinetic was analyzed by quantifying the paramter adhesion time [A_T]. This term is defined as the time needed until half of the starting cell density at time 0 h, adhere to material surface.

$$N = N_0 * e^{(-A_R * t)} \qquad A_R = \frac{\ln 2}{A_T}$$

Figure 10: Formulas for calculating the kinetic of adhesion mechanism
adhesion time [A_T], adhesion rate [A_R]; N number of nonadherent cells at time t = n; N_0 number of nonadherent cells at time 0 h; t time [h].

This measurement was performed with all used cell types cultivated on control samples, silicone elastomer, the hydrogels HESHEMA and PEG and on adhesion ligand-coated slides placed in a petri dish with a diameter of 35 mm filled with 2 ml culture media. After cultivation times of 1, 2, 3, 4 and 5 h or 10, 20, 30, 40, 50 min for the ligand substrates, respectively the culture medium (including all of the nonadherent cells) of each petri dish was collected and served as the cell suspension going to be counted using a Fuchs Rosenthal cell counter device. For a better comparison between the experiments, the cell density was normalized on the seeding density at time 0 h and given in percent. The cell densities were used to calculate adhesion time [A_T] (Figure 10). Every result was averaged over four independent measurements ± SEM.

4.4.4 Microscopic analysis

Several optical methods were applied to visualize the cells cultivated on the samples. Independently from observing the cells during cultivation time in general or on three-dimensional features with the help of a light microscope (Nikon, Düsseldorf, Germany), the following procedures were used. All results were quantified with ImageJ software (http://rsbweb.nih.gov/ij/).

Adhesion pattern

To analyze adhesion pattern of cells, the surface reflectance interference contrast (SRIC) technique was used. By reflecting light at the interface between the cell and the cultivation surface, SRIC microscopy allows the visualization of focal contacts. The closer the adhesion contacts between the cell and the surface, the darker appear interference fringes. SRIC analysis is limited to transparent surfaces, therefore, the adhesion pattern only on glass samples and adhesion ligand-coated slides were investigated. After 5 h or 24 h cultivation time the cells were fixed with 4 % formaldehyde solved in PBS for 10 min and conserved in PBS. Images were recorded using a fluorescence microscope equipped with a SRIC filter set (Nikon TE 2000-E, Nikon, Düsseldorf, Germany) and a CCD camera. The software "NIS Elements AR 3.0" (Nikon, Düsseldorf, Germany) was used to acquire the images.

For quantification the ImageJ software was applied. With the help of an area selection tool, the cell area shown in the recorded images was surrounded manually. Afterwards, a histogram was created which displays the distribution of gray values with a scale from 0 (pure black) to 255 (pure white) in the active selection. The possible gray values correlate with the distance between the cells and the interface referring to SRIC technique. Automatically, the relative number of pixels found for each gray value was counted and mean, standard derivation, minimum and maximum were calculated. The results were averaged over at least 100 cells per treatment coming from four independent measurements. Adhesion pattern of human fibroblasts, GM-7373 endothelial cells, SH-SY5Y neuroblastoma cells, HaCaT keratinocytes, MG-63 osteoblasts, and A10 smooth muscle cells were analyzed under control conditions and in dependence of adhesion ligands with maximum concentration after the shortterm and longterm experimental setup.

Investigation of cell morphology

Cell morphology was analyzed by fluorescence after nucleus and actin filaments staining using 4', 6-diamidino-2-phenylindole-dihydrochlorid:hydrat (DAPI) and phalloidin-Alexa 488, respectively (Molecular probes Invitrogen, Grenzach-Whylen, Germany). After 5 h and 24 h cultivation time, cells grown on the samples were fixed by a 10 min incubation in PBS containing 4 % formaldehyde. Then the cells were permeabilized by incubation in PBS containing 0.3 % Triton X-100 for 10 min. The chromatin in nucleus was stained by an incubation in PBS containing 1 μM DAPI for

10 min. After washing with PBS, actin filaments were stained with 0.6 U phalloidin-Alexa 488 solved in PBS in the dark for 1 h. For further analysis the cells were conserved in PBS.

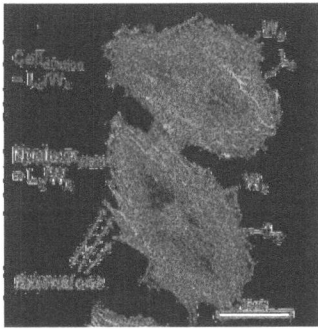

Figure 11: Quantification of cell morphology

The morphology of the cells was observed with a fluorescence microscope (Nikon TE 2000-E, Nikon, Düsseldorf, Germany) at excitation lights of 348 nm for DAPI and 488 nm for phalloidin-Alexa. Images were acquired using a CCD camera and software "E Z-C1 3.5" (Nikon, Düsseldorf, Germany). The quantitative evaluation of the results was performed with ImageJ software (Figure 11). First, length and width of each single nucleus (L_n, W_n) and each single cell (L_c, W_c) were measured. As the scales are automatically given in pixels, each length and width was converted into μm. By calculating the quotient L_n/W_n and L_c/W_c the nucleus and cell dilation were estimated. Second, the number of cell extensions such as filopodia, lamellipodia and retraction fibres was counted which were defined as appendages that taper off to the surface and to neighboring cells. The results were given as mean ± SEM for four independent measurements. About 100 cells per treatment were evaluated. This procedure was performed with all cell types under control conditions and on the adhesion ligand-coated substrates after the shortterm and longterm experimental setup, but also with laser-fabricated surface structures.

Cell orientation

To investigate the effects of different grating structures produced in titanium on cell orientation, human fibroblasts, and MG-63 osteoblasts were stained and images were created following the description in 'Investigation of cell morpholgy' after 24 h cultivation time. With the help of a line selection tool using ImageJ software, a straight line was placed on each single cell over the total cell length. Automatically, an angle for each line was recorded that refers to the cell orientation within the recorded image. The results were given as the standard derivation of the averaged angles of at least 200 cells per treatment. The decrease of parallel orientation of the cells correlates with an increase of the calculated standard derivation.

4.4.5 Proliferation assay

Biomaterial-cell interaction can also be characterized via analyzing material effects on cell growth. For this purpose, proliferation profiles of all used cell types cultivated after the laser-induced forward procedure and on the adhesion ligand-coated slides after the longterm experimental setup, on polymers, hydrogels, and metals also in dependence of material chemistry and surface topographies were examined.

After different times of cultivation the adherent cells were trypsinized. To determine the cell density, the cell suspension was collected and centrifuged at 800 g for 10 min. The cells in pellet were resolved in cell culture media and counted using a Fuchs Rosenthal cell counter device or the cell counter Casy TT$^®$ from Innovatis (Bielefeld, Germany). For a better comparison between the experiments, the cell densities were normalized in percent on the seeding density at time 0 h. Furthermore, the doubling time [h] was calculated defined as the time needed for passing once the cell cycle. The results were given as average ± SEM of four independent experiments.

4.4.6 Analysis of gap junction coupling

The formation of cells to a real functioning tissue requires gap junction coupling. For characterizing gap junction coupling the so-called scrape loading method was used. By scratching the monolayer of cells and adding lucifer yellow, this fluorescent dye can penetrate in the destroyed cells. As gap junction channels are permeable for lucifer yellow, the diffusion distance over channel-connected neighboring cells as a sign for gap junction coupling can be estimated. Experimental conditions of the scrape loading method followed previous description [108]. The investigations were performed with the adhesion ligand-coated slides in comparison to the control.

GM-7373 endothelial cells, human fibroblasts, HaCaT keratinocytes, and A10 smooth muscle cells were cultivated on adhesion coated glass slides with different concentrations and on the control within a 24-well plate including 2 ml culture media. After 24 h cultivation time a monolayer was formed. Then the slides were carefully washed with NaCl-BS for 2 min. Afterwards the slides were placed in a NaCl-BS solution including 0.25 % lucifer yellow (Sigma-Aldrich, Taufkirchen, Germany) and with the help of a razor blade two straight scratches along the whole sample were set. After 5 min incubation time the slides were washed twice with NaCl-BS for 5 min each. At last the cells were fixed by a 10 min incubation in PBS containing 4 % formaldehyde and conserved in PBS.

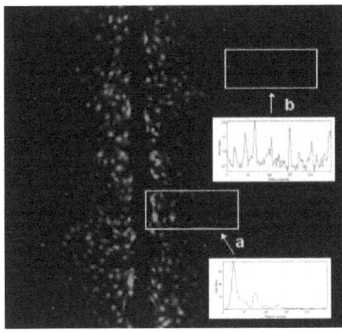

Figure 12: Quantification of gap junction coupling showing rectangles (250 x 100 px) and plot profiles of (a) diffusion distance of lucifer yellow and (b) background

The images (1024 x 1024 [px]) of each scratch were recorded using a fluorescence microscope (Nikon TE 2000-E, Nikon, Düsseldorf, Germany) at an excitation light of 488 nm and a CCD camera using the software "E Z-C1 3.5" (Nikon, Düsseldorf, Germany). With the help of ImageJ software (http://rsbweb.nih.gov/ij/) for each image six rectangles with the dimension of 250 x 100 [px] were placed along the scratch including the start of the scratch and diffusion area of lucifer yellow. One rectangle aside the scratch was applied to measure the background signal. After that a plot profile was used to display a two-dimensional graph of the intensities of pixels along a line within each selected rectangle (Figure 12). The x-axis represents the horizontal distance through the selection and the y-axis the vertically averaged pixel intensity. For each treatment the six plot profiles coming from 16 separate images were averaged minus each background signal. The results were given as average of the diffusion distance of lucifer yellow ± SEM.

4.5 Statistical analysis

In comparison to the control treatments Student's-t-test analysis (two-sided, independent, $p < 0.05$, 0.01 and 0.001) was used to estimate statistical significant differences.

5 Results

5.1 Cell responses to unstructured materials

5.1.1 Materials influenced DNA strand breaking

DNA damage effects were characterized by comet assay and quantified with the parameter tailmoment. On the control treatment GFSHR-17 granulosa cells showed a tailmoment of 2.56 ± 048. This value was comparable when cultivated on Ormocomp® and PEG SR610 (2 wt% photoinitiator Bis), with tailmoments of 2.64 ± 0.41 and 2.72 ± 0.77, respectively. On the contrary, on PEG SR259 (2 wt% photoinitiator Bis) the tailmoment was significantly increased to 23.34 ± 4.2. The presence of Ormocomp® and PEG SR610 did not significantly increase the incidence of DNA damage effects, whereas PEG SR259 did.

	Tailmoment ± SEM		
Cell type	Fibroblasts	Endothelial	Neuroblastoma
Control	1.48 ± 0.13	1.72 ± 0.1	1.25 ± 0.1
PEG SR610 + 2 % PI 2959 fresh	2.98 ± 0.56***	6.81 ± 0.58***	3.42 ± 0.49***
PEG SR610 + 2 % PI 2959 aged	1.66 ± 0.39	1.62 ± 0.23	1.23 ± 0.17

Table 1: Analysis of DNA damage effects of PEG SR610 (2 wt% photoinitiator Irgacure 2959) in dependence of material aging demostrated by comet assay of human fibroblasts, GM-7373 endothelial cells, and SH-SY5Y neuroblastoma cells.
The results were given as average of tailmoment as a marker of DNA damages ± SEM of four independent measurements after 24 h cultivation time. At least 1000 cells per treatment were evaluated.
*** statistical difference ($p < 0.001$) in comparison to the control via Students-t-test analysis.

On the control surface human fibroblasts, GM-7373 endothelial, and SH-SY5Y neuroblastoma cells showed a tailmoment of 1.48 ± 0.13, 1.72 ± 0.1, and 1.25 ± 0.1, respectively (Table 1). On fresh PEG SR610 (2 wt% photoinitiator Irgacure 2959) the tailmoments were significantly increased to 2.98 ± 0.56, 6.81 ± 0.58, and 3.42 ± 0.49 (Table 1). After material aging, DNA damages were comparable with the control with tailmoments of 1.66 ± 0.39, 1.62 ± 0.23, and 1.23 ± 0.17, respectively (Table 1). Whereas fresh PEG samples increased significantly the incidence of DNA damage effects, on aged samples it was decreased for all investigated cell types.

5.1.2 Materials affected adhesion time in a cell specific manner

Adhesion kinetic was quantified by the parameter adhesion time A_T. The results were normalized to the starting cell densities [cell/ml] of $1.75*10^6$ for human fibroblasts, of $1.44*10^6$ for GM-7373 endothelial cells, and of $6.37*10^6$ for SH-SY5Y neuroblastoma cells. On the control surface adhesion times A_T [h] of 2.52 ± 0.19 for fibroblasts, 4.11 ± 0.73 for endothelial cells, and 3.79 ± 0.62 for neuroblastoma cells were found (Table 2). On silicone elastomer the adhesion time A_T [h] of fibroblasts was significantly increased to 16.34 ± 1.52. Simultaneously, the adhesion times A_T [h] of endothelial cells and neuroblastoma cells were reduced to 2.02 ± 0.04 and 2.27 ± 0.4, respectively (Table 2). On HESHEMA (DS 0.11) and aged PEG SR610 (2 wt% photoinitiator Irgacure 2959) fibroblasts adhered faster with adhesion times A_T [h] of 1.83 ± 0.41 and 1.34 ± 0.06. On the contrary, the adhesion times A_T [h] of endothelial cells were significantly increased to 24.85 ± 5.41 and 5.01 ± 1.39, respectively. The adhesion times A_T [h] of neuroblastoma cells were significantly increased to 8.19 ± 2.31 on HESHEMA and to 5.99 ± 2.14 on PEG (Table 2).

	Adhesion time A_T [h] ± SEM			
Cell type	Control	Silicone	HESHEMA	PEG
Fibroblasts	2.52 ± 0.19	16.34 ± 1.52 *	1.83 ± 0.41	1.34 ± 0.06
Endothelial	4.11 ± 0.73	2.02 ± 0.04	24.85 ± 5.41 *	5.01 ± 1.39
Neuroblastoma	3.79 ± 0.62	2.27 ± 0.4	8.19 ± 2.31 *	5.99 ± 2.14

Table 2: Adhesion time A_T [h] of human fibroblasts, GM-7373 endothelial cells, and SH-SY5Y neuroblastoma cells on silicone elastomer, HESHEMA (DS 0.11), and aged PEG SR610 (2 wt% photoinitiator Irgacure 2959).
In comparison to the control over 5 h cultivation time; the results were presented as average ± SEM of four independent measurements, referring to the seeding cell density at t = 0 h ($1.75*10^6$, $1.44*10^6$, and $6.37*10^6$ cells/ml, respectively).
* statistical difference (p < 0.05) in comparison to the control via Student's-t-test analysis.

5.1.3 Materials influenced proliferation in a cell specific manner

Concerning Ormocomp®, the cell densities of GFSHR-17 granulosa cells were determined after 8, 24, 32, and 48 h, of GM-7373 endothelial and SH-SY5Y neuroblastoma cells after 24, 48, 72, and 96 h cultivation time. Each proliferation measurement was started with an average cell density [cells/ml] of $5.76*10^6$, $3.65*10^6$, and $4.63*10^6$, respectively and normalized in percent.

Results

In terms of quantity, granulosa cells reached 687.5 % ± 40.3 on the control surface after 48 h cultivation time. On Ormocomp® the cell density was 848.96 % ± 27.7 (Figure 13 a). After 96 h cultivation time endothelial cells proliferated to 1093.49 % ± 16.31 and neuroblastoma cells to 2862.56 % ± 19.6 under control conditions. The proliferation was comparable when cultivated on Ormocomp® with cell densities [%] of 1113.97 ± 17.29 and 2500 ± 18.7, respectively (Figure 13 a, b).

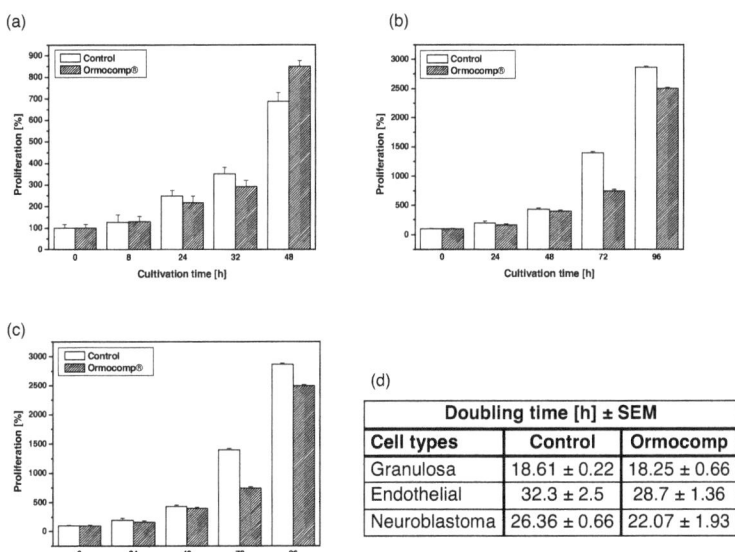

Figure 13: Proliferation profiles of (a) GFSHR-17 granulosa cells, (b) GM-7373 endothelial cells, and (c) SH-SY5Y neuroblastoma cells, (d) doubling times [h] on polymer Ormocomp® in comparison to the control over 48 h or 96 h cultivation time.
The results were normalized on the starting cell density (5.76*10^6, 3.65*10^6, and 4.63*10^6 cells/ml, respectively) and given as average (in percent) ± SEM of four independent measurements.

Under control conditions granulosa cells showed a doubling time [h] of 18.61 ± 0.22, endothelial cells of 28.7 ± 1.36, and neuroblastoma cells of 22.07 ± 1.93. In the presence of Ormocomp® comparable doubling times [h] of 18.25 ± 0.66, 28.7 ± 1.36, and 22.07 ± 1.93 were achieved (Figure 13 d). Each cell type cultivated on Ormocomp® grew as fast as under control conditions.

On silicone elastomer, three different HESHEMA derivatives (DS 0.07, 0.11, and 0.2), and PEG (SR610, 2 wt% photoinitiator Irgacure 2959) in dependence of material aging cell growth of human fibroblasts, GM-7373 endothelial cells, and SH-SY5Y neuroblastoma cells were examined after 8, 24, 32 and 48 h cultivation time. The cell densities were given in percent normalized on the starting density [cells/ml] of $1.21*10^6$, $4.68*10^6$, and $1.47*10^6$, respectively.

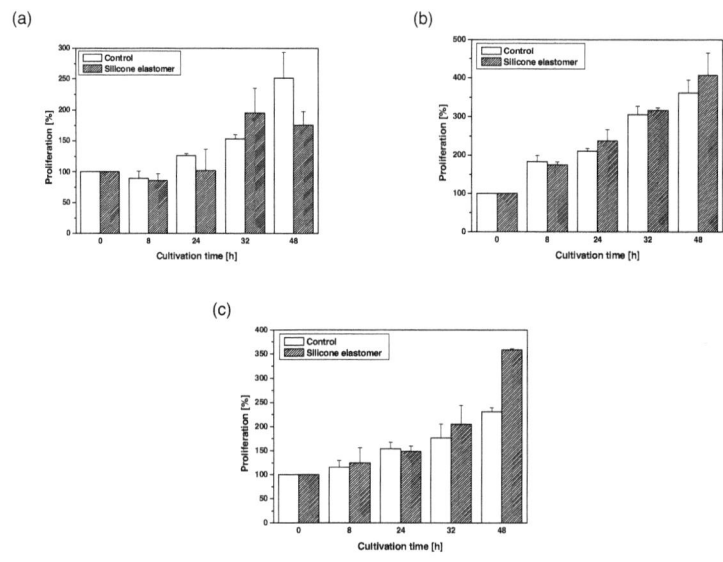

Figure 14: Proliferation profiles of (a) human fibroblasts, (b) GM-7373 endothelial cells, and (c) SH-SY5Y neuroblastoma cells on polymer silicone elastomer in comparison to the control over 48 h cultivation time.
The results were presented as average ± SEM of four independent measurements, normalized in percent on the seeding cell density at t = 0 h ($1.21*10^6$, $4.68*10^6$, and $1.47*10^6$ cells/ml, respectively).

On the control surface, fibroblasts reached a cell density [%] of 251.21 ± 42.52, endothelial cells of 361.65 ± 33.75, and neuroblastoma cells of 230.86 ± 8.4 after 48 h cultivation time. On silicone elastomer fibroblasts decreased the cell density [%] to 175.22 ± 21.88 (Figure 14 a). Endothelial cells grew at the same rate as on the control with 407.38 ± 57.77 (Figure 14 b). Neuroblastoma cells showed a tendency to accelerate their proliferation to 358.62 ± 2.43 when cultivated on silicone elastomer (Figure 14 c).

After 48 h cultivation time under control conditions, cell densities [%] of 251.21 ± 42.53 for fibroblast, 361.65 ± 33.75 for endothelial, and 230.86 ± 8.4 for neuroblastoma cells were found. On HESHEMA derivatives in the order of 0.07, 0.11 and 0.2 DS fibroblasts proliferated comparably to 202.12 ± 14.9, 213.34 ± 40 and 239.07 ± 14.66 (Figure 15 a). Endothelial cells reduced the growth to 45.5 ± 7.18, 84.49 ± 31.98, and 99.76 ± 10.04, respectively (Figure 15 b). Similarly to endothelial cells, neuroblastoma cells reached 25.57 ± 13.24, 28.93 ± 16.9 and 111.82 ± 12 on HESHEMA (Figure 15 c).

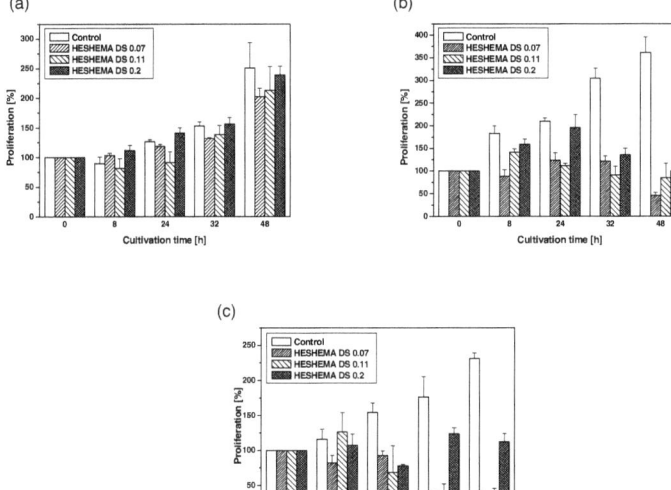

Figure 15: Proliferation profiles of (a) human fibroblasts, (b) GM-7373 endothelial cells, and (c) SH-SY5Y neuroblastoma cells on hydrogel HESHEMA in dependence of the DS-value (0.07, 0.11, 0.2) in comparison to the control over 48 h cultivation time.
The results were given as average ± SEM of four independent measurements, normalized in percent on the seeding cell density at $t = 0$ h ($1.21*10^6$, $4.68*10^6$, and $1.47*10^6$ cells/ml, respectively).

After 48 h cultivation time fibroblasts, endothelial and neuroblastoma cells showed cell densities [%] of 271 ± 39.42, 361.65 ± 33.75, and 230.86 ± 8.4 on the control surface. On fresh PEG samples the proliferation was reduced to 102.04 ± 38.22, 45.95 ± 28.07 and 47.23 ± 8.36, respectively (Figure 16). On aged PEG samples fibroblasts proliferated at the same rate as under control conditions up to 299.54 ± 1.95 (Figure 16 a). Endothelial cells and neuroblastoma cells reduced their cell growth [%] to 62.4 ± 13.78 (Figure 16 b) and 116.2 ± 24.49 (Figure 16 c).

(a)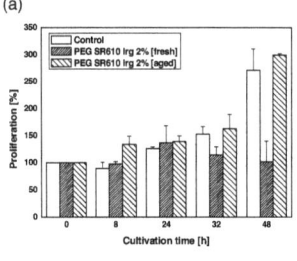

(b)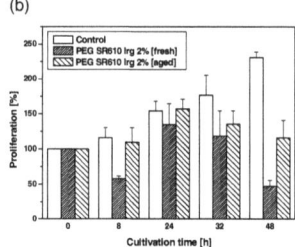

(c)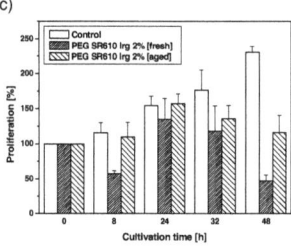

Figure 16: Proliferation profiles of (a) human fibroblasts, (b) GM-7373 endothelial cells, and (c) SH-SY5Y neuroblastoma cells on hydrogel PEG SR610 (2 wt% photoinitiator Irgacure 2959) in dependence of material aging in comparison to the control over 48 h cultivation time.
The results were given as average ± SEM of four independent measurements, normalized in percent on the seeding cell density at t = 0 h (7.39*10^6, 1.81*10^6, and 1.15*10^6 cells/ml, respectively).

5.2 Cell responses to three-dimensional scaffolds

5.2.1 Scaffolds composed of Ormocomp® and PEG SR610

Three-dimensional scaffolds in micrometer scale were produced by two-photon polymerization technique by Dr. A. Ovsianikov (Laser Zentrum Hannover e. V., Germany).

Ormocomp® was used to fabricate two types scaffolds placed on glass slides of 18 x 18 mm. The first type was a periodic grating structure with a total area of 1 mm² and a height of < 5 μm. Line distances of each square varied from 10, 20, 30, 40 to 50 μm. The second type were cylinders with a height of 100 μm and an average diameter of 10 - 100 μm.

Scaffolds composed of PEG SR610 consisted of rings that were arranged next to each other without spacing. Not only different diameters of these rings but also varying numbers of ring layers were produced on a total area of 1 mm². The height of the scaffold was arranged between 100 - 200 µm. Scanning electron microscopy (SEM) images are shown in Figure 17.

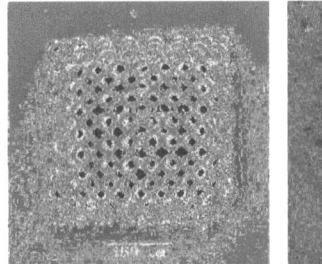 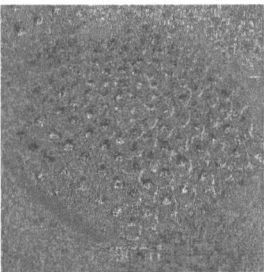

Figure 17: SEM images of laser-fabricated three-dimensional scaffolds composed of PEG SR610.
The samples were produced and pictured by Dr. A. Ovsianikov at the Laser Zentrum Hannover e. V. (Germany)

5.2.2 Microscopic analysis of different cell types on three-dimensional scaffolds

Because of the small size dimensions, human fibroblasts were not visible by light microscopy on Ormocomp® gratings with the size dimensions of 10 µm and 20 µm. In Figure 18 cells images are shown of 30, 40, and 50 µm gratings recorded after 4 and 10 days of cultivation. Independent from the grating size, it was observed that after 4 days cultivation time fibroblasts fell into the grating squares and adapted their morphology to the feature dimensions. Moreover, it was found that the cells were able to proliferate over the total cultivation time. Nevertheless, morphological differences were seen after 10 days. On 30 µm and 40 µm gratings fibroblasts also adhered on the top of the gratings acquiring their normal elongated cell shape, even though they were rather placed within the gratings of 30 µm (Figure 18 b, d). On the contrary, the cells did not adhere on the top of 50 µm gratings (Figure 18 f).

Results

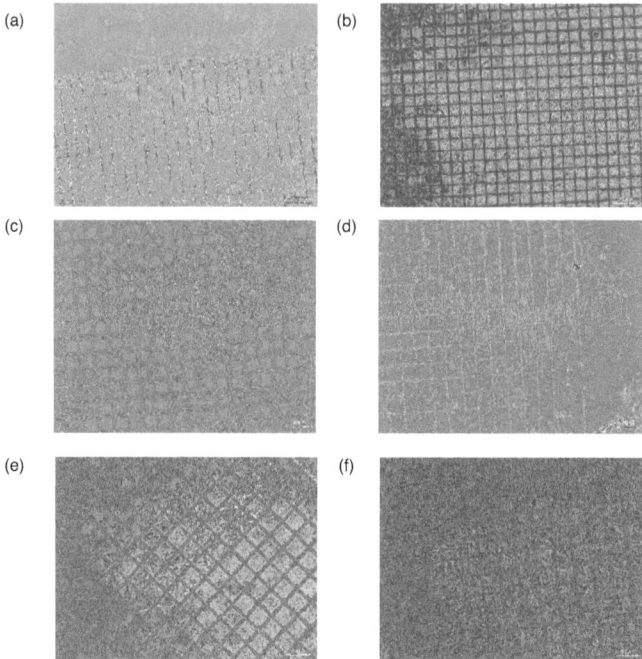

Figure 18: Microscopic images of human fibroblasts cultivated on laser-fabricated grating structures of Ormocomp® in dependence of grating size.
(a) 30 µm grating after 4 days, (b) 30 µm grating after 10 days, (c) 40 µm grating after 4 days, (d) 40 µm grating after 10 days, (e) 50 µm grating after 4 days, (f) 50 µm grating after 10 days in culture.
The structures were generated at the Laser Zentrum Hannover e. V. (Germany).

Whether SH-SY5Y neuroblastoma cells adhere on lateral surfaces, was investigated with cylindrical structures composed of Ormocomp® over a total cultivation time of 4 days.

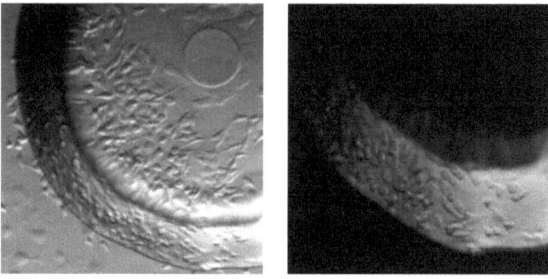

Figure 19: Stereo microscope images of SH-SY5Y neuroblastoma cells cultivated on laser-fabricated cylinders composed of Ormocomp®.
The structures were generated by Dr. A. Ovsianikov at the Laser Zentrum Hannover e. V. (Germany).

To avoid a sedimentation of the cells, they were kept in suspension by the use of a shaking table (Figure 8). The microscopic analysis of SH-SY5Y neuroblastoma cells revealed that the cells were able to adhere on lateral surfaces and form layers, which spread from the bottom to the top of the three-dimensional structures (Figure 19).

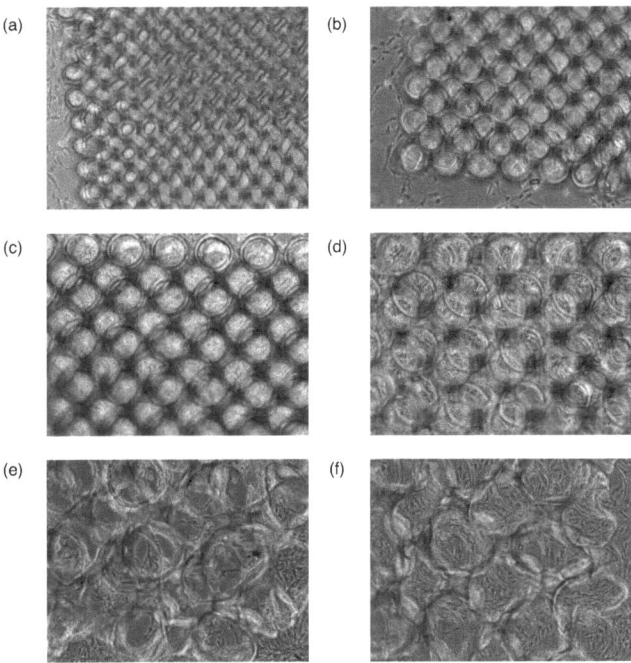

Figure 20: Microscopic images of GM-7373 endothelial cells on laser-fabricated PEG SR610 scaffolds.
With varying diameters of (a) 60 μm, (b) 80 μm, (c) 100 μm, (d) 120 μm, (e) 180 μm, and (f) 200 μm after 4 days cultivation time.
The structures were generated by Dr. A. Ovsianikov at the Laser Zentrum Hannover e. V. (Germany).

On PEG SR610 scaffolds with a diameter of 60 μm GM-7373 endothelial cells were hardly visible (Figure 20 a). Similarly to fibroblasts, endothelial cells fell within the features with varying diameters of 80, 100, 120, 180, and 200 μm. The images in Figure 20 show that scaffold size independently the cells fell down on the bottom of the substrate, adhered, presented a normal elongated shape, and proliferated after 4 days cultivation time.

Results

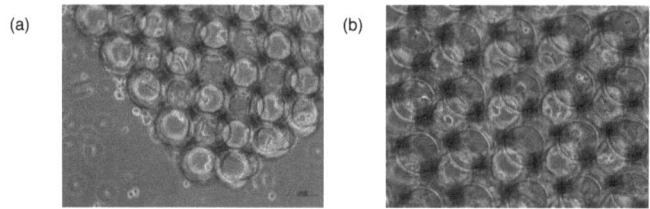

Figure 21: Microscopic images of NIH3T3 fibroblasts on laser-fabricated PEG SR610 scaffolds with varying diameters of (a) 50 μm and (b) 70 μm after 4 days cultivation time.
The structures were generated by Dr. A. Ovsianikov at the Laser Zentrum Hannover e. V. (Germany).

PEG SR610 scaffolds with diameters of 50 μm and 70 μm were used for NIH3T3 fibroblasts. It was observed that the cells fell within the features size independently and presented a rounded shape after 4 days cultivation time (Figure 21).

5.3 Cell transport with laser-induced forward transfer

5.3.1 Arrangement of cells in defined pattern

With the laser-induced forward transfer NIH3T3 fibroblasts and HaCaT keratinocytes were transferred and arranged in a two-dimensional chess-like pattern. The pattern with a total size of 9.6 x 9.6 mm consisted of four lines per square with a line width of 70 μm and a line spacing of 200 μm. Fibroblasts were dyed green and keratinocytes blue by using CFDA and Hoechst 33342.

Figure 22: Fluorescence image of NIH3T3 fibroblasts (bright gray) and HaCaT keratinocytes (dark gray) arranged by laser-induced forward transfer.
The image was produced by Dipl.-Ing. M. Grüne and Dr. L. Koch at the Laser Zentrum Hannover e. V. (Germany).

The image in Figure 22, produced by Dipl.-Ing. M. Grüne and Dr. L. Koch at the Laser Zentrum Hannover e. V. (Germany), shows the precise creation of pattern with more than one cell type produced with the laser-induced forward transfer.

5.3.2 Analysis of DNA damage effects after laser-induced forward transfer

DNA damage effects were characterized by comet assay and quantified with the parameter tailmoment. On the control fibroblasts showed a tailmoment of 1.81 ± 0.33, keratinocytes of 2.41 ± 0.38, human mesenchymal stem cells of 1.27 ± 0.28, and porcine mesenchymal stem cells of 1.33 ± 0.15. After the transport with the laser-induced forwar transfer the tailmoments were in the same range of 1.72 ± 0.43, 2.31 ± 0.25, 1.32 ± 0.28, and 1.18 ± 0.13, respectively (Table 3).

	Tailmoment ± SEM	
Cell type	Control	LIFT
Fibroblasts	1.81 ± 0.33	1.72 ± 0.43
Keratinocytes	2.41 ± 0.38	2.31 ± 0.25
Human mesenchymal stem	1.27 ± 0.28	1.32 ± 0.28
Porcine mesenchymal stem	1.33 ± 0.15	1.18 ± 0.13

Table 3: Analysis of DNA damage effects after laser-induced forward transfer of NIH3T3 fibroblasts, HaCaT keratinocytes, human mesenchymal stem cells, porcine mesenchymal stem cells in comparison to the control demonstrated by comet assay.
The results were given as average of tailmoment, a marker of DNA fragmentation, ± SEM of at least 1000 cells per treatment.

5.3.3 Cell proliferation after laser-induced forward transfer

Cell growth of NIH3T3 fibroblasts and HaCaT keratinocytes was documented after 24, 48, 72, and 96 h cultivation time right after the transfer and under control conditions. Each treatment was started with an average cell density of 10^3 cells/ml and normalized in percent.

After 96 h cultivation time cell densities [%] of 904.69 ± 182.22 for fibroblasts and of 912.28 ± 130.64 for keratinocytes were found under control conditions. After the transfer the proliferation was comparable with 927.19 ± 183.34 and 1028.46 ± 31, respectively (Figure 23). This measurement was performed by Dipl.-Ing. M. Grüne and Dr. L. Koch at the Laser Zentrum Hannover e. V. (Germany).

(a) 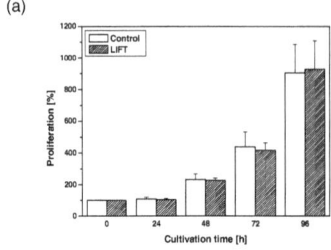 (b)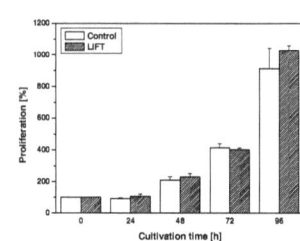

Figure 23: Proliferation profiles of (a) NIH3T3 fibroblasts and (b) HaCaT keratinocytes after laser-induced forward transfer (LIFT) and under control conditions over 96 h cultivation time. The results were given as average ± SEM of six independent measurements, normalized in percent on the starting cell density of 10^3 cells/ml at time t = 0 h.

5.4 Cell responses to laser-fabricated surface topographies

5.4.1 Surface topographies for material functionalization

Surface structuring was accomplished by material ablation with femtosecond lasers and negative replication process performed by Dipl.-Phys. E. Fadeeva at the Laser Zentrum Hannover e. V. (Germany). By changing the laser processing parameters different structure features with variable size dimensions were produced.

So-called spike structures in silicon and silicone elastomer were observed with a scanning electron microscope (SEM) and presented an array of quasi-periodical conical spikes. By adjusting the laser fluence from 0.36 J/cm² to 3.6 J/cm² the spike to spike distance of silicon was changed from 2 µm to 15 µm and the spike height from 1 µm to 20 µm (Figure 24).

Figure 24: SEM images of laser-fabricated spike structures in silicon at different laser fluences [J/cm²].
Structuring and imaging were performed by Dipl.-Phys. E. Fadeeva at the Laser Zentrum Hannover e. V. (Germany).

For the cell experiments silicon spike structures with an average height of 5.9 μm and a spike to spike distance of 4.8 μm were produced. The average top flattening of the spikes was 1.4 μm in diameter.

Furthermore, silicon spikes were used for negative replication process to transfer and reproduce the structures in silicone elastomer (Figure 25).

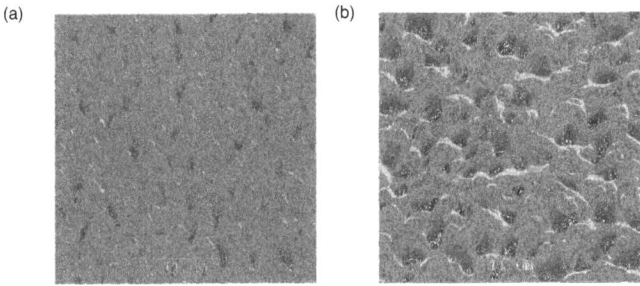

Figure 25: SEM images of (a) laser-fabricated silicon spikes and (b) the negative replicas in silicone elastomer.
Structuring and imaging were performed by Dipl.-Phys. E. Fadeeva at the Laser Zentrum Hannover e. V. (Germany).

Titanium was used to produce two different types of surface topographies on a total surface area of 3 x 3 mm. The first type consisted of a hierarchical nano- and micro-superimposed structure, which was self-organized and randomly orientated. Scanning electron microscopy (SEM) images in Figure 26 demonstrate that these microstructures also represented a nano-roughness from the bottom to the top of each spike.

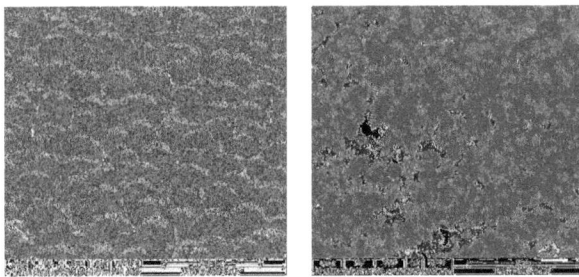

Figure 26: SEM images of hierarchical nano- and microsuperimposed surface structures in titanium with different magnifications.
Structuring and imaging were performed by Dipl.-Phys. E. Fadeeva at the Laser Zentrum Hannover e. V. (Germany).

Results

The second surface type in titanium represented a periodic groove structure. Size dimensions of the groove structures are shown as a histogram (Figure 27). Groove width were varied between 5, 10, 15, 20, 25, and 30 µm. The depth of the structures was ≥ 2 µm.

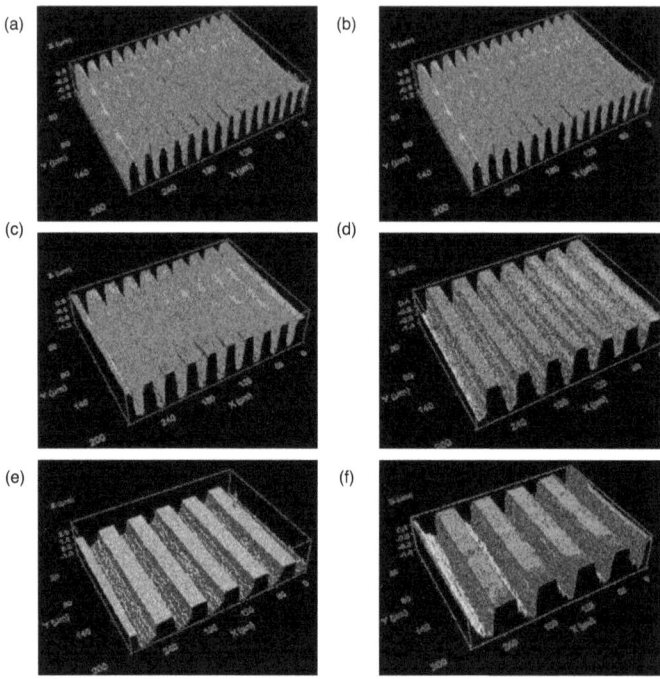

Figure 27: Histograms of groove structures in titanium in dependence of groove width. (a) 5 µm, (b) 10 µm, (c) 15 µm, (d) 20 µm, (e) 25 µm, and (f) 30 µm.
Structuring and imaging were performed by Dipl.-Phys. E. Fadeeva at the Laser Zentrum Hannover e. V. (Germany).

In platinum 24 different structure features with variable size dimensions in nanometer scale were produced. Exemplarily shown are surface roughness consisting of nano- and microscale cavities, nanoprotrusions and microscale aggregates (Figure 28 a - c) and groove structures (Figure 28 d - e).

Results

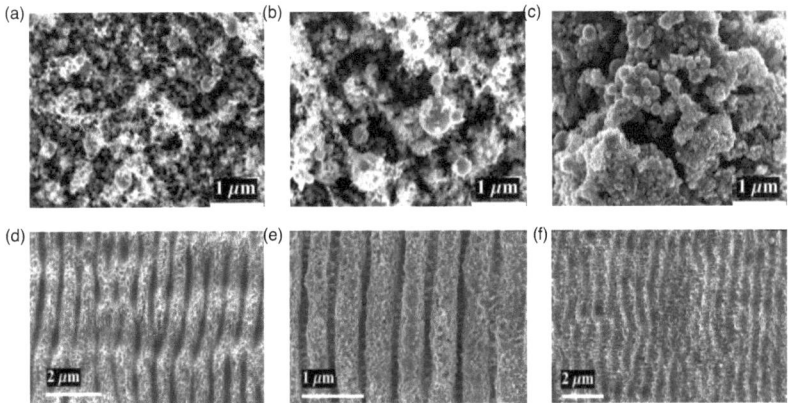

Figure 28: SEM images of laser-fabricated nanostructures in platinum at different laser processing parameters. (a - c) nanoroughness and (d - e) nanogrooves.
The structures were prepared by A. Y. Vorobyev (University of Rochester, USA).

5.4.2 Topography induced wettability changes

The analysis of material chemistry revealed that surface structuring did not change the elemental compositions of the materials. The oxid layers were removed by material washing with hydrofluoric acid (HF). Both methods were performed by Dipl.-Phys. E. Fadeeva at the Laser Zentrum Hannover e. V. (Germany).

Material	Water contact angle [°] ± 3 SEM		Structure type
	unstructured	structured	
Silicon	62	130	micrometer spikes
Silicone elastomer	130	159	
Titanium	80	160	hierarchical
Platinum	78	110 - 158	nanoroughness
		85 - 129	nanogrooves

Table 4: Water contact angle measurements of silicon, silicone elastomer, titanium, and platinum in dependence of surface structuring.
This measurement was performed by Dipl.-Phys. E. Fadeeva at the Laser Zentrum Hannover e. V. (Germany).

The sessile drop method was used to investigate material wettability which was performed by Dipl.-Phys. E. Fadeeva at the Laser Zentrum Hannover e. V. (Germany). Unstructured silicon, silicone elastomer, titanium, and platinum presented a water contact angle [°] of 62, 130, 80, and 78, respectively (Table 4). After fabricating the spike structures, the angles [°] were increased to 130 for silicon and to 159 for silicone elastomer. On the structures in titanium the angle was increased to 160°. On platinum surfaces presenting a nanoroughness the contact angles [°] were arranged between 110 and 158, on the nanogroove features between 85 and 129 (Table 4).

5.4.3 Topography influenced DNA strand breaking

DNA damage effects were characterized by comet assay and quantified with the parameter tailmoment. On the control human fibroblasts and SH-SY5Y neuroblastoma cells showed a tailmoment of 1.47 ± 0.13 and 1.25 ± 0.1, respectively (Table 5). On unstructured materials such as silicon, silicone elastomer, titanium, and platinum the tailmoments of fibroblasts were in a comparable range of 1.4 ± 0.34, 1.53 ± 0.5, 1.45 ± 0.55, and 1.32 ± 0.24, respectively (Table 5). Neuroblastoma cells presented a tailmoment of 1.24 ± 0.19 on silicon and of 1.13 ± 0.28 on silicone elastomer (Table 5).

Material	Structure	Tailmoment ± SEM	
		Fibroblasts	Neuroblastoma
Control	unstructured	1.47 ± 0.13	1.25 ± 0.1
Silicon	unstructured	1.4 ± 0.34	1.24 ± 0.19
	spike	1.48 ± 0.52	1.1 ± 0.17
Silicone elastomer	unstructured	1.53 ± 0.5	1.13 ± 0.28
	spike	2.6 ± 0.52 ***	1.86 ± 0.86 ***
	spike (negative replica)	1.3 ± 0.33	1.45 ± 0.25
Titanium	unstructured	1.45 ± 0.55	-
	hierarchical	1.48 ± 0.42	
Platinum	unstructured	1.32 ± 0.24	-
	nanostructures 0.5 J/cm^2	1.16 ± 0.19	
	nanostructures 1.5 J/cm^2	1.21 ± 0.2	
	nanostructures 3.5 J/cm^2	0.9 ± 0.2	

Table 5: Analysis of DNA damage effects of surface topographies in silicon, silicone elastomer, titanium, and platinum demonstrated by comet assay of human fibroblasts and SH-SY5Y neuroblastoma cells in comparison to controls after 24 h cultivation time.
The results were given as average of tailmoment as a marker of DNA damages ± SEM of four independent measurements. At least 1000 cells per treatment were evaluated.
*** statistical difference ($p < 0.001$) in comparison to the control via Students-t-test analysis.

On directly ablated spike structures in silicone elastomer the tailmoments of both cell types were significantly increased. For fibroblasts a tailmoment of 2.6 ± 0.52 and for neuroblastoma cells of 1.86 ± 0.86 were found (Table 5). Comparable with the control treatment neuroblastoma cells presented tailmoments of 1.24 ± 0.19 on silicon structures and 1.45 ± 0.25 on spike replicas in silicone elastomer (Table 5). This result was also found for fibroblasts with tailmoments of 1.48 ± 0.52 and 1.3 ± 0.33, respectively (Table 5). On the hierarchical nano- and micro- superimposed structures in titanium fibroblasts also showed a comparable tailmoment of 1.48 ± 0.42 (Table 5). None of the nanostructures produced in platinum with different laser processing parameters such as laser fluence [J/cm^2] increased the incidence of DNA damage effects for fibroblasts. The cells showed tailmoments of 1.16 ± 0.19, 1.21 ± 0.2, and 0.9 ± 0.2, respectively (Table 5).

5.4.4 Topograhical effects on orientation and cell morphology

The effects on cell orientation of groove structures produced in titanium were analyzed in dependence of groove width and cell type such as human fibroblasts and MG-63 osteoblasts. Parallel orientation was quantified by the standard derivation of the averaged cell orientation within the images. Fibroblasts reduced the parallel orientation on groove width larger than 15 μm (Figure 29 a - f, Table 6). Concerning osteoblasts, a reduced parallel orientation began on widths larger than 25 μm (Figure 29 e - l, Table 6).

Groove width [μm]	Standard derivation of orientation	
	Fibroblasts	Osteoblasts
5	5.3	3.01
10	8.35	4.83
15	14.38	4.01
20	14.54	5.93
25	15.13	16.03
30	21.32	27.29

Table 6: Quantification of cell orientation of human fibroblasts and MG-63 osteoblasts on groove structures of titanium in dependence of groove width after 24 h cultivation time.
Groove width varied between 5, 10, 15, 20, 25, and 30 μm. The results were given as standard derivation of orientation of at least 200 cells per treatment.

Results

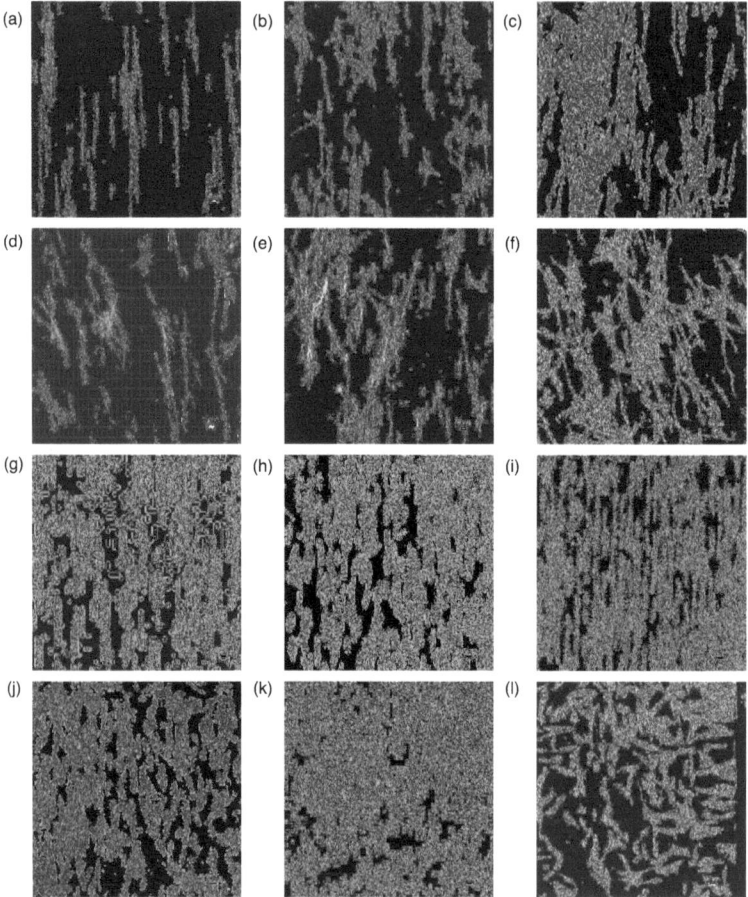

Figure 29: Fluorescence images (stained actin filaments) of human fibroblasts (a - f) and MG-63 osteoblasts (g - l) cultivated on groove structures in titanium in dependence of groove width after 24 h cultivation time.
(a, g) 5 µm, (b, h) 10 µm, (c, i) 15 µm, (d, j) 20 µm, (e, k) 25 µm, and (f, l) 30 µm.

The hierarchical nano- and micro- superimposed structures in titanium were used to analyze topographical effects on cell morphology of human fibroblasts and MG-63 osteoblasts. On the control fibroblasts were elongated (Figure 30 a). On the structures the cells were rounded (Figure 30 b). No significant differences for osteoblasts were observed when cultivated on the control and on the structures (Figure 30 c, d).

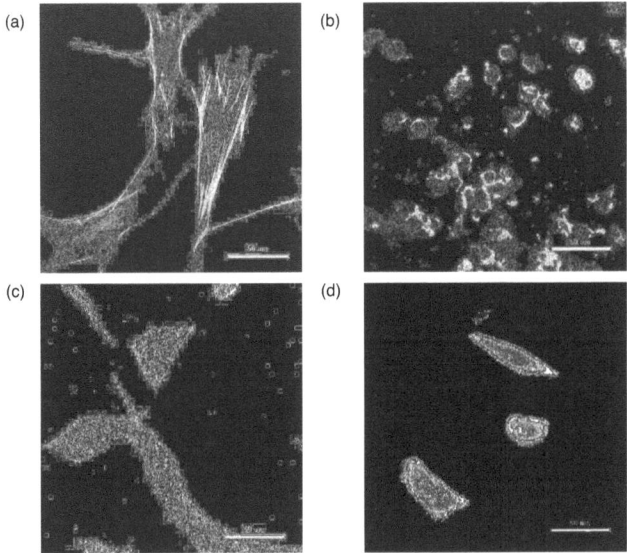

Figure 30: Fluorescence images of human fibroblasts on (a) control and (b) hierarchical titanium structures, and of MG-63 osteoblasts on (c) control and (d) hierarchical titanium structures after 24 h cultivation time.

Topographical effects on cell morphology of human fibroblasts and SH-SY5Y neuroblastoma cells were tested with spike structures in silicon. The results were quantified by nucleus and cell dilation given as the ratio of L_n/W_n and L_c/W_c. On the control fibroblasts showed a nucleus dilation of 1.59 ± 0.04. L_n/W_n was similar on silicon unstructured with 1.69 ± 0.1. On the spike structures L_n/W_n was significantly increased to 1.98 ± 0.11 (Figure 31 a). For neuroblastoma cells it was found that the control nucleus dilation of 2.34 ± 0.09 was significantly decreased on unstructured silicon to 1.68 ± 0.07 and to 1.48 ± 0.04 on the spikes (Figure 31 a). Under control conditions the cell dilation of fibroblasts with 6.37 ± 0.41 was significantly decreased to 4.34 ± 032 on unstructured silicon and to 2.92 ± 0.26 structured silicon (Figure 31 a). No difference in L_c/W_c of neuroblastoma cells occurred on the control surface and unstructured silicon with comparable L_c/W_c of 5.77 ± 0.39 and 4.97 ± 0.34. On the spikes their cell dilation was significantly decreased to 2.96 ± 0.23 (Figure 31 a).

Results

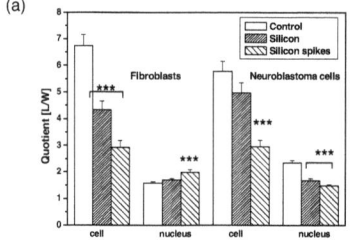 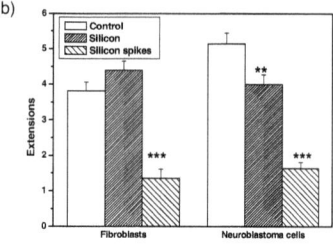

Figure 31: Quantification of cell morphology of human fibroblasts and SH-SY5Y neuroblastoma cells cultivated on silicon and silicon spikes in comparison to the control after 24 h cultivation time.
(a) Quotient [L/W] of nucleus and cell, (b) number of extensions. The results were given as average ± SEM of at least 100 cells per treatment.
*** statistical difference (p < 0.001) in comparison to the control via Student's-t-test analysis.

Moreover, the average number of extensions was evaluated. On the control surface fibroblasts formed 3.82 ± 0.24 and neuroblastoma cells 5.14 ± 0.29 extensions (Figure 31 b). On unstructured silicon the average number of 4.34 ± 0.25 for fibroblasts was comparable with the control. On the spikes fibroblasts significantly reduced the formation of extensions to 1.35 ± 0.26 (Figure 31 b). Neuroblastoma cells formed significantly less extensions on unstructured silicon and on structured silicon. The number of extensions were 4 ± 0.26 and 1.63 ± 0.18, respectively (Figure 31 b).

5.4.5 Topography affected proliferation in a cell specific manner

The proliferation of human fibroblasts and SH-SY5Y neuroblastoma cells grown on silicon spikes and the negative replica in silicone elastomer were studied after 48 h cultivation time. The results were normalized in percent on the starting cell density of $1.6*10^6$ cells/ml. On the control and on unstructured silicon cell densities [%] of 283.32 ± 57.74 and 290.36 ± 22.08 were found for fibroblasts (Table 7). On unstructured silicone elastomer, on silicon spikes and on silicone elastomer spikes fibroblasts reduced the cell densities [%] to 224.01 ± 8.96, 155.52 ± 14.9, and significantly to 95.95 ± 12.15, respectively (Table 7). On the control the cell density [%] of 230.86 ± 47.23 was reproduced on all tested materials and surface structures for neuroblastoma cells (Table 7). Comparable cell densities [%] of 257.12 ± 52.2 on unstructured silicon, 251.62 ± 28.05 on silicon spikes, 243.42 ± 26.03 on silicone elastomer unstructured, and 258.91 ± 28.76 on spike replicas in silicone elastomer were found (Table 7).

Cell type	Proliferation [%] ± SEM	
	Fibroblasts	Neuroblastoma
Control	283.32 ± 57.74	230.86 ± 47.23
Silicon unstructured	290.36 ± 22.08	257.12 ± 52.2
Silicon spikes	155.52 ± 14.9	251.62 ± 28.05
Silicone elastomer unstructured	224.01 ± 8.96	243.42 ± 26.03
Silicone elastomer spikes (negative replicas)	95.94 ± 12.15 *	258.91 ± 28.76

Table 7: Proliferation results of human fibroblasts and SH-SY5Y neuroblastoma cells cultivated on unstructured and structured silicon and silicone elastomer in comparison to the control after 48 h.
The results were given as average ± SEM of four independent measurements, normalized on the starting cell densities of $1.6*10^6$ cells/ml at t = 0 h.
* statistical difference ($p < 0.05$) in comparison to the control via Student's-t-test analysis.

The effects of hierarchical nano- and micro- superimposed structures in titanium on cell proliferation were tested with human fibroblasts and MG-63 osteoblasts. The cell densities were normalized in percent on the starting cell densities of either $9.7*10^4$ or $1.13*10^5$ cells/ml.

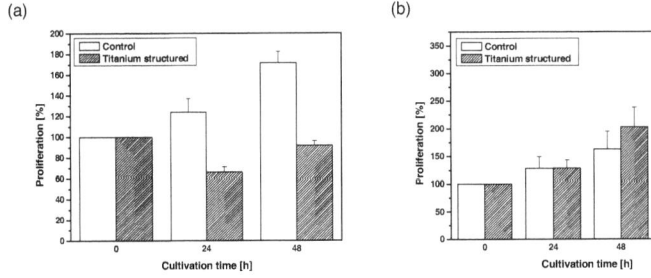

Figure 32: Proliferation profiles of (a) human fibroblasts and (b) MG-63 osteoblasts cultivated on hierarchical nano- and micro- superimposed titanium structures and under control conditions over 48 and 72 h cultivation time.
The results were given as average ± SEM of four independent measurements, normalized in percent on the starting cell density at time t = 0 h ($9.7*10^4$ and $1.13*10^5$ cells/ml, respectively).

Under control conditions a cell density [%] of 171.64 ± 10.9 was found for fibroblasts. On the structures fibroblasted proliferated to 91.93 % ± 4.5 after 48 h cultivation time (Figure 32 a). On the control surface osteoblasts presented a cell density [%] of

299.59 ± 18.95, on the structures of 331.72 ± 19.89 after 72 h cultivation time (Figure 32 b).

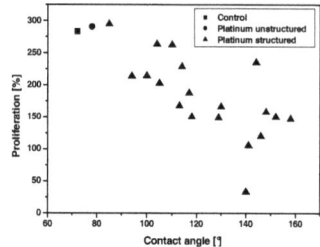

Figure 33: Proliferation profiles of human fibroblasts cultivated on nanostructured platinum and under control conditions for 48 h cultivation time.
The results were normalized in percent on the starting cell density at time t = 0 h (1.3*10^6 cells/ml).

Cell growth of human fibroblasts cultivated on nanostructured platinum was examined after 48 h cultivation time, normalized on the starting cell density of 1.3*10^6 cells/ml. On unstructured platinum fibroblasts proliferated in a similar manner as under control conditions to 283.32 % and 290.63 %. Cultivated on structured samples the proliferation of fibroblasts was reduced. The decreasing wetting of the surfaces was in accord with a more significant reduction of cell growth (Figure 33).

5.5 Analysis of adhesion kinetic and pattern

5.5.1 Cell specific adhesion kinetic

Adhesion kinetic was quantified by the parameter adhesion time A_T. This parameter is normalized on the starting cell densities [cells/ml] of 9.17*10^5 for human fibroblasts, 2.36*10^6 for GM-7373 endothelial cells, for 3.3*10^6 SH-SY5Y neuroblastoma cells, for 2.68*10^7 HaCaT keratinocytes, 1.1*10^6 for MG-63 osteoblasts, and 3.3*10^6 for A10 smooth muscle cells.

In Table 8 it is shown that the cells attached to the control surface with a specific speed, characterized by adhesion time A_T [h]. Following ranking was found: osteoblasts with 1.84 ± 0.09 > keratinocytes with 2.51 ± 0.18 ≅ fibroblasts with 2.52 ± 0.19 >

neuroblastoma cells with 3.79 ± 0.62 > endothelial cells with 4.11 ± 0.73 > smooth muscle cells with 4.54 ± 0.78.

Cell type	Adhesion time A_T [h] ± SEM
Fibroblasts	2.52 ± 0.19
Endothelial cells	4.11 ± 0.73
Neuroblastoma cells	3.79 ± 0.62
Keratinocytes	2.51 ± 0.18
Osteoblasts	1.84 ± 0.09
Smooth muscle cells	4.54 ± 0.78

Table 8: Adhesion time A_T [h] of human fibroblasts, GM-7373 endothelial cells, SH-SY5Y neuroblastoma cells, HaCaT keratinocytes, MG-63 osteoblasts, and A10 smooth muscle cells under control conditions.
The results were presented as average ± SEM of four independent measurements.

5.5.2 Cell specific adhesion pattern

After 24 h cultivation time under control conditions the adhesion pattern of human fibroblasts, GM-7373 endothelial cells, SH-SY5Y neuroblastoma cells, HaCaT keratinocytes, MG-63 osteoblasts, and A10 smooth muscle cells were examined.

Neuroblastoma cells (Figure 34 c), keratinocytes (Figure 34 d), and osteoblasts (Figure 34 e) rather formed many small focal contacts to the control surface. On the contrary, fibroblasts (Figure 34 a), endothelial cells (Figure 34 b), and smooth muscle cells (Figure 34 f) showed also wide areas of close contacts and seemed to adhere with the whole cell body.

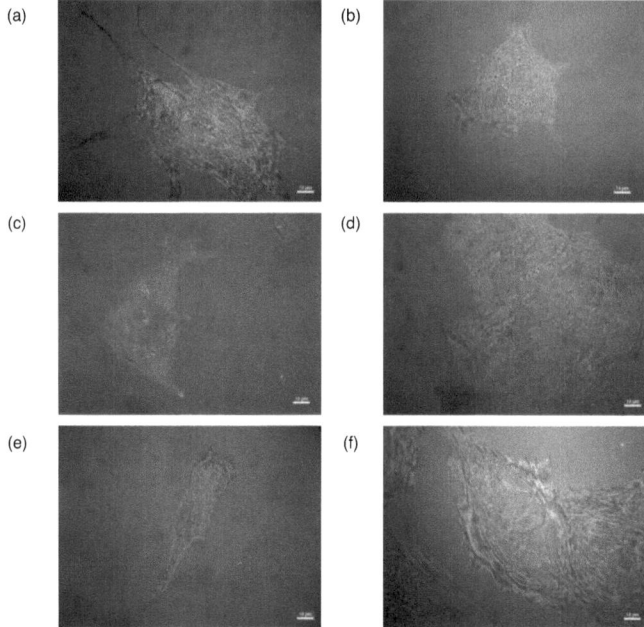

Figure 34: Control adhesion pattern via SRIC-technique of (a) human fibroblasts, (b) GM-7373 endothelial cells, (c) SH-SY5Y neuroblastoma cells, (d) HaCaT keratinocytes, (e) MG-63 osteoblasts, and (f) A10 smooth muscle cells after 24 h cultivation time.

5.6 Cell responses to adhesion ligands

5.6.1 Shortterm effects of adhesion ligands

The measurements were restricted to a cultivation time of 5 h in serum-free cell culture media with maximum ligand concentrations [$\mu g/cm^2$] such as 2 for laminin, 5 for fibronectin, 10 for collagen and 0.1 for vitronectin.

Adhesion ligands affected adhesion pattern in a cell specific manner

On the control surface all cell types formed many small focal contacts instead of attaching with the whole cell body (figures not shown). The adhesion pattern of keratinocytes were not affected by the ligands. On the following treatments, the cells attached with their whole cell body: fibroblasts on fibronectin, collagen, and laminin, endothelial cells and osteoblasts on vitronectin, and neuroblastoma cells on laminin. In contrast to all the other investigated cell types, smooth muscle cells still formed wide sections of close contacts to the ligand surfaces. But differences occurred with respect to contact localization. Whereas on vitronectin the sections were restricted to the outer cell area, on laminin, fibronectin and collagen sections in the inner cell part were also found. This pattern was most pronounced when cultivated on laminin.

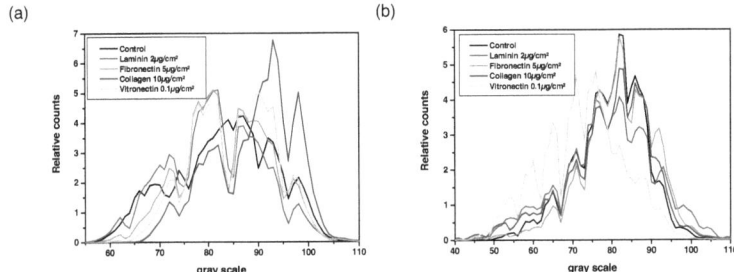

Figure 35: Histogram of adhesion pattern of (a) SH-SY5Y neuroblastoma cells and (b) MG-63 osteoblasts on adhesion ligands after 5 h cultivation time.
Histogram of gray values which correlates with the distance between the cells and the surface. The results were given as average of the relative distribution of gray values within each cell body of at least 100 cells per treatment.

To get more insight into the ligand effects on adhesion pattern, the images were quantified via a histrogram that represents the distribution of each gray value of the cell bodies. Two histograms are exemplarily shown in Figure 35. For a better comparison between the treatments, the average ± SEM, minimum and maximum of each gray value distribution were examined.

		Control	Laminin 2 µg/cm²	Fibronectin 5 µg/cm²	Collagen 10 µg/cm²	Vitronectin 0.1 µg/cm²
Fibroblasts	Minimum	21	27	27	20	11
	Maximum	188	147	166	182	195
	Average ± SEM	78.04 ± 3.26	75.48 ± 2.69 *	75.48 ± 2.5	79.12 ± 2.99	79.97 ± 2.98
Endothelial	Minimum	35	20	30	29	19
	Maximum	111	178	120	160	130
	Average ± SEM	76.12 ± 2.43	73.48 ± 3.31	77.42 ± 2.42	77.86 ± 2.81	74.98 ± 3.12
Neuro-blastoma	Minimum	51	46	49	61	55
	Maximum	116	128	137	127	141
	Average ± SEM	83.15 ± 1.58	80.25 ± 1.75 *	83.73 ± 1.56	88.87 ± 1.14	84.96 ± 1.87
Keratino-cytes	Minimum	24	10	23	14	26
	Maximum	160	151	153	148	134
	Average ± SEM	79.69 ± 2.24	77.46 ± 3.07	81.49 ± 2.46	79.91 ± 3.05	72.69 ± 2.14 *
Osteoblasts	Minimum	42	37	37	56	41
	Maximum	118	119	115	123	144
	Average ± SEM	79.98 ± 1.88	81.2 ± 1.84	78.08 ± 1.96	82.34 ± 1.66	82.31 ± 2.11
Smooth muscle	Minimum	36	38	44	48	45
	Maximum	122	142	158	175	159
	Average ± SEM	88.37 ± 2.77	86.72 ± 2.91	86.34 ± 3.18	89.1 ± 2.61	87.4 ± 2.98

Table 9: Quantification of adhesion pattern of human fibroblasts, GM-7373 endothelial cells, SH-SY5Y neuroblastoma cells, HaCaT keratinocytes, MG-63 osteoblasts, and A10 smooth muscle cells in dependence of adhesion ligands after 5 h cultivation time.
The results were given as minimum, maximum and average ± SEM of gray values within each cell body of at least 100 cells per treatment.
* statistical difference ($p < 0.05$) in comparison to the control via Students-t-test analysis.

No significant differences were found for fibroblasts, when cultivated on the control surface, fibronectin, collagen and vitronectin. All investigated values were in the same range with a minimum to maximum of about 20 to 190, and an average gray value of about 78. On the contrary, laminin effects were significant. The maximum and average values were reduced to 147 and 75, respectively (Table 9).

Under control conditions, on fibronectin, and collagen minimum values of about 29 -35 were found for endothelial cells. On laminin and vitronectin the minimum value was reduced to about 20. The maximum value followed the order laminin with 178 > collagen with 160 > vitronectin with 130 > fibronectin with 120 > control with 111. For the control surface, fibronectin, and collagen the average gray values were comparable with about 77, but they were decreased to 74 on laminin and vitronectin (Table 9).

In contrast to fibroblasts and endothelial cells, the minimum values for neuroblastoma cells were switched to a larger cell-surface distance starting at 46 for laminin followed by fibronectin, control, vitronectin and collagen. On the control surface the maximum value was 116, on collagen 127, on laminin 128, on fibronectin 137, and on vitronectin 141. Under control conditions an average value of 83.15 ± 1.58 was found. On fibronectin, collagen, and vitronectin this value was in the same range. It was significantly decreased to 80.25 ± 1.75 on laminin (Table 9).

Except for vitronectin with an average gray value of 72.69 ± 2.14, the other treatments had comparable values of about 80 for keratinocytes. Concerning the minimum values, the treatments could be separated into two groups. The first group consisting of laminin and collagen formed closer cell-surface distances at about 10. For the second group including the control surface, fibronectin, and vitronectin the closest distance was transferred to values of about 24. On the contrary, the maximum values followed a ranking of vitronectin with 134, collagen with 148, laminin with 151, fibronectin with 153, and the control surface with 160 (Table 9).

With respect to the examined gray values of osteoblasts, the control, laminin, and fibronectin were comparable. The similar distributions of about 40 to 119 were increased on collagen from 56 to 123 and on vitronectin from 41 to 144. Simultaneously, the average gray values did not show significant differences for all treatments and were in a range of 78 to 82 (Table 9).

Under control conditions and on laminin, smooth muscle cells showed minimum gray values of 36 and 38. Cultivated on fibronectin, collagen, and vitronectin the contact distances started at 44 to 48. The maximum values followed the order collagen with 175 > vitronectin ≅ fibronectin with 159 and 158 > laminin with 142 > control surface with 122. For all treatments the average gray value was in the same range of 86 to 89 (Table 9).

Adhesion ligands affected cell morphology in a cell specific manner
Exemplarily, the images of SH-SY5Y neuroblastoma cells are shown (Figure 36). On the control, on fibronectin, and on collagen neuroblastoma cells presented a rounded shape and a small number of extensions (Figure 36 a, c, d). On laminin and vitronectin the cells were elongated and increased the formation of extensions (Figure 36 b, e).

Results

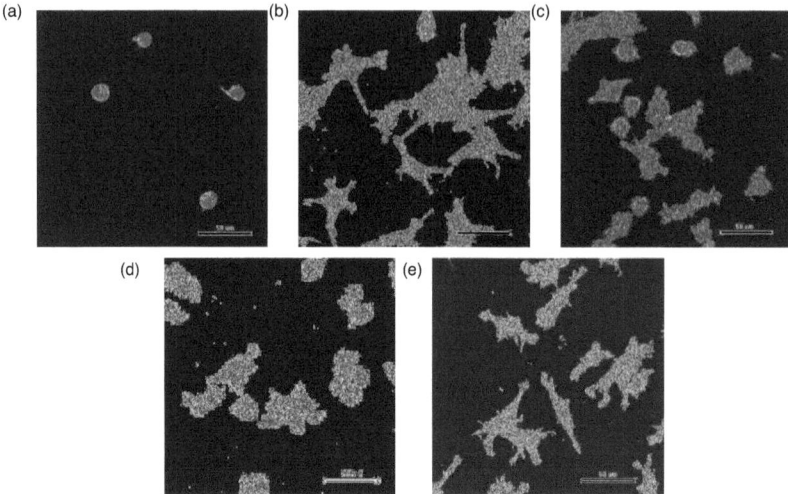

Figure 36: Fluorescence images of SH-SY5Y neuroblastoma cells on (a) control, (b) laminin, (c) fibronectin, (d) collagen, and (e) vitronectin after 5 h cultivation time.

To quantify adhesion ligands effects on cell morphology, the nucleus and cell dilation were examined and given as the ratios of L_n/W_n and L_c/W_c. Furthermore, the average number of extensions was determined.

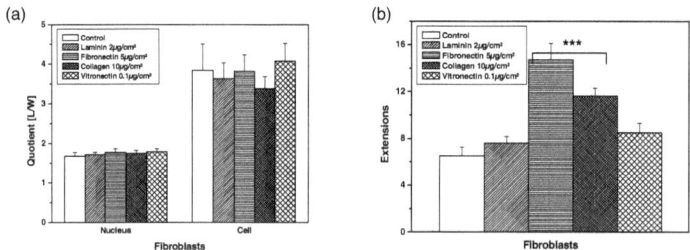

Figure 37: Quantification of cell morphology of human fibroblasts cultivated on adhesion ligands in comparison to the control after 5 h cultivation time.
(a) Quotient [L/W] of nucleus and cell, (b) number of extensions. The results were given as average ± SEM of at least 100 cells per treatment.
*** statistical difference ($p < 0.05; 0.01; 0.001$) in comparison to the control via Students-t-test analysis.

For fibroblasts nucleus and cell dilations on the ligands were comparable with the control of 1.68 ± 0.09 and 3.84 ± 0.66, respectively (Figure 37 a). Concerning the average number of extensions, 6.52 ± 0.74 were found under control conditions,

7.63 ± 0.53 on laminin, 8.53 ± 0.79 on vitronectin. Significantly more extensions were formed on collagen with 11.63 ± 0.64 and on fibronectin with 14.71 ± 1.38 (Figure 37 b).

Endothelial cells presented comparable nucleus dilation of 1.51 ± 0.56 and cell dilation of 2.84 ± 0.55 on all treatments (Figure 38 a). Under control conditions the cells formed 3.53 ± 0.53 extensions. On fibronectin and collagen the number of extensions were comparable with 2.43 ± 0.19 and 3.63 ± 0.23 (Figure 38 b). On laminin and especially vitronectin the formation of extensions was significantly increased to 5.87 ± 0.37 and 6.97 ± 0.42, respectively (Figure 38 b).

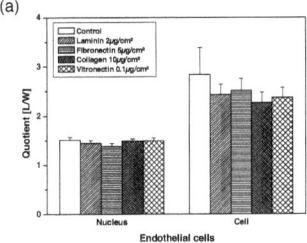 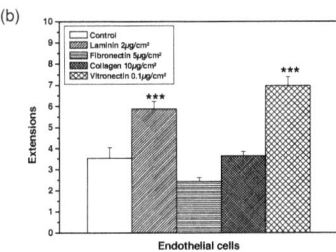

Figure 38: Quantification of cell morphology of GM-7373 endothelial cells cultivated on adhesion ligands in comparison to the control after 5 h cultivation time.
(a) Quotient [L/W] of nucleus and cell, (b) number of extensions. The results were given as average ± SEM of at least 100 cells per treatment.
*** statistical difference ($p < 0.05; 0.01; 0.001$) in comparison to the control via Students-t-test analysis.

The fluorescence images of neuroblastoma cells demonstrated rounded cell shapes on the control surface, fibronectin, and collagen (Figure 36 a, c, d). The analysis of cell dilation given as L_c/W_c revealed comparable values of about 1.8 on these treatments. On laminin and vitronectin the cells were more elongated with L_c/W_c of 4.11 ± 0.4 and 3.62 ± 0.29 (Figure 39 a). All nucleus dilations were comparable and ranged about 1.5 (Figure 39 a). Furthermore, the average number of extensions reflected similarities between the control, fibronectin, and collagen with values of about 1.8 (Figure 39 b). On laminin and vitronectin a significant increase of average extensions to 6.47 ± 0.36 and 5.23 ± 0.26 was found (Figure 39 b).

(a) 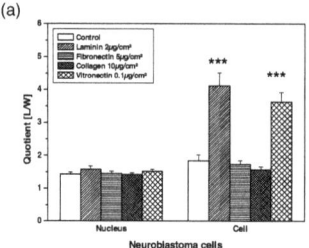 (b)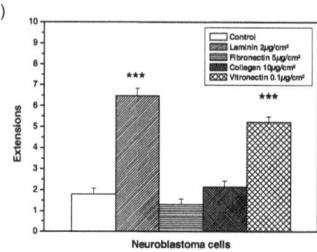

Figure 39: Quantification of cell morphology of SH-SY5Y neuroblastoma cells cultivated on adhesion ligands in comparison to the control after 5 h cultivation time.
(a) Quotient [L/W] of nucleus and cell, (b) number of extensions. The results were given as average ± SEM of at least 100 cells per treatment.
*** statistical difference (p < 0.05; 0.01; 0.001) in comparison to the control via Students-t-test analysis.

L_n/W_n of keratinocytes' treatments were comparable and ranged in about 1.4. On collagen it was significantly increased to 1.83 ± 0.1 (Figure 40 a). A cell dilation of 1.54 ± 0.11 was found on the control surface, on laminin and vitronectin. On collagen and fibronectin L_c/W_c was 3.73 ± 0.3 and 2.99 ± 0.22, respectively (Figure 40 a). The number of extensions was similar between the control surface of 2.77 ± 0.48 and on laminin and vitronectin. On the contrary, more extensions were formed on fibronectin and collagen with about 8 (Figure 40 b).

(a) 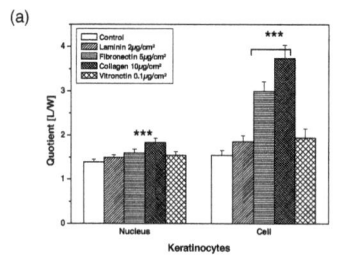 (b)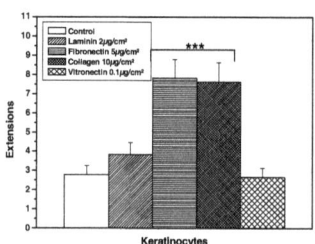

Figure 40: Quantification of cell morphology of HaCaT keratinocytes cultivated on adhesion ligands in comparison to the control after 5 h cultivation time.
(a) Quotient [L/W] of nucleus and cell, (b) number of extensions. The results were given as average ± SEM of at least 100 cells per treatment.
*** statistical difference (p < 0.05; 0.01; 0.001) in comparison to the control via Students-t-test analysis.

All adhesion ligands did not affect the nucleus dilation of osteoblasts and had values of about 1.6 (Figure 41 a). The L_c/W_c of 3.62 ± 0.48 on the control surface was also found on laminin and fibronectin. It was significantly increased to 5.51 ± 0.59 on collagen and to 5 ± 0.46 on vitronectin (Figure 41 a). With respect to the average number of extensions, on the control surface 5.07 ± 0.64 extensions were formed. This number was significantly increased on vitronectin to 8.53 ± 0.85. On collagen the number of extensions was also higher than the control with 6.2 ± 0.61, but not significantly. The values for laminin and fibronectin were in the same range like under control conditions (Figure 41 b).

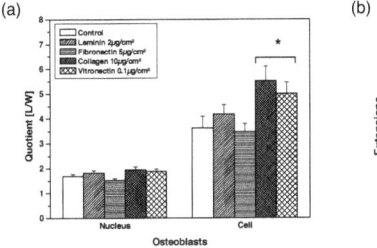
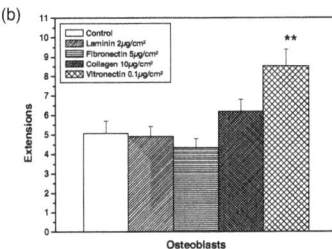

Figure 41: Quantification of cell morphology of MG-63 osteoblasts cultivated on adhesion ligands in comparison to the control after 5 h cultivation time.
(a) Quotient [L/W] of nucleus and cell, (b) number of extensions. The results were given as average ± SEM of at least 100 cells per treatment.
*** statistical difference ($p < 0.05$; 0.01; 0.001) in comparison to the control via Students-t-test analysis.

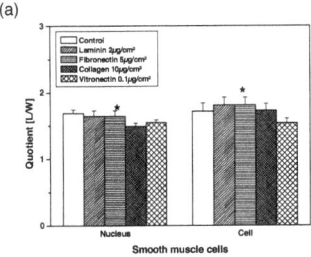
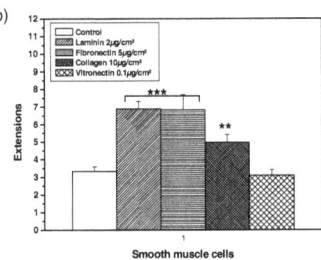

Figure 42: Quantification of cell morphology of A10 smooth muscle cells cultivated on adhesion ligands in comparison to the control after 5 h cultivation time.
(a) Quotient [L/W] of nucleus and cell, (b) number of extensions. The results were given as average ± SEM of at least 100 cells per treatment.
*** statistical difference ($p < 0.05$; 0.01; 0.001) in comparison to the control via Students-t-test analysis.

On the control a nucleus dilation of 1.69 ± 0.06 and a cell dilation of 1.72 ± 0.12 was found for smooth muscle cells. Most of the adhesion ligands did not affect both ratios except for collagen and laminin (Figure 42 a). A significant decrease of L_n/W_n was observed on collagen to 1.49 ± 0.05, a significant increase of L_c/W_c on laminin to 1.81 ± 0.12 (Figure 42 a). The number of extensions followed the ranking laminin and fibronectin with about 6.9 > collagen with 4.97 ± 0.44 > control and vitronectin with about 3.4. On vitronectin the effect was not significant (Figure 42 b).

5.6.2 Longterm effects of adhesion ligands

The longterm measurements were performed with serum-containing cell culture media in dependence of the ligand concentration.

Adhesion ligands accelerated adhesion kinetic

Exemplarily, the adhesion profiles of human fibroblasts, GM-7373 endothelial cells, and SH-SY5Y neuroblastoma cells on the maximum ligand concentrations are shown,

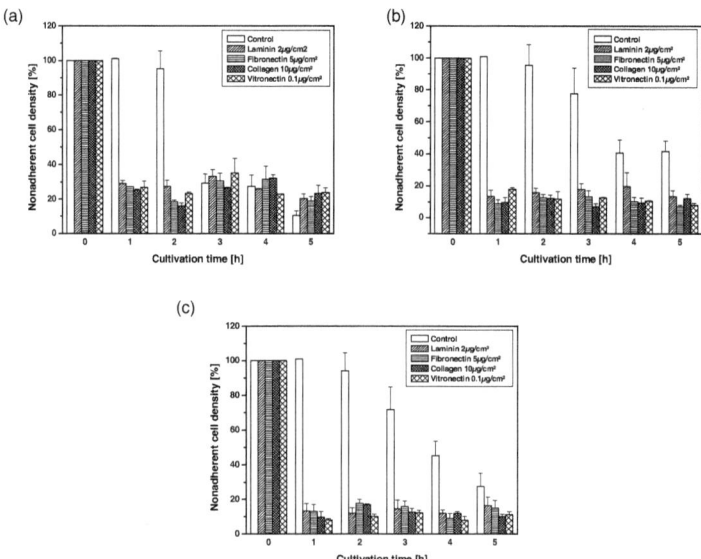

Figure 43: Adhesion profiles of (a) human fibroblasts, (b) GM-7373 endothelial cells, and (c) SH-SY5Y neuroblastoma cells on adhesion ligands over 5 h cultivation time.
In comparison to the control; the results were given as average $\pm$ SEM of four independent measurements, normalized in percent on the seeding cell density at t = 0 h ($5.88*10^5$, $1.11*10^5$, and $1.41*10^5$ cells/ml, respectively).

normalized on the starting cell density [cells/ml] of $5.88*10^5$, $1.11*10^5$, and $1.41*10^5$, respectively.

As shown in Figure 43, after 1 h cultivation about 70 % of fibroblasts, about 80 % of endothelial cells, and about 90 % of neuroblastoma cells adhered on the ligand surfaces. Similar amounts of adhesion were found after 5 h cultivation under control conditions. Cell type and ligand independently, adhesion kinetic was accelerated, reflected by the adhesion time A_T.

Adhesion ligands influenced adhesion kinetic in a cell specific manner

Adhesion kinetic on the maximum ligand concentrations was quantified by the parameter adhesion time A_T. This parameter is normalized on the starting cell densities [cells/ml] of $1.17*10^5$ for human fibroblasts, of $1.26*10^5$ for GM-7373 endothelial cells, of $1.06*10^5$ for SH-SY5Y neuroblastoma cells, of $2.05*10^5$ for HaCaT keratinocytes, of $1.85*10^5$ for MG-63 osteoblasts, and of $2.2*10^5$ for A10 smooth muscle cells.

Cell types	Adhesion time A_T [min] ± SEM					
	Fibro-blasts	Endo-thelial	Neuro-blastoma	Keratino-cytes	Osteo-blasts	Smooth muscle
Laminin 2 [µg/cm²]	24.23 ± 2.67	17.03 ± 1.28	16.89 ± 1.99	55.02 ± 5.56	20.04 ± 0.9	25.18 ± 2.14
Laminin 1 [µg/cm²]	22.15 ± 3.43	18.03 ± 0.63	17.48 ± 0.9	64.35 ± 3.49	25.62 ± 1.82	27.29 ± 3.17
Fibronectin 5 [µg/cm²]	15.89 ± 1.34	25.89 ± 1.79	17.57 ± 0.71	44.99 ± 3.17	20.56 ± 1.24	25.37 ± 1.38
Fibronectin 3 [µg/cm²]	21.6 ± 1.39	21.19 ± 1.92	17.59 ± 1.32	56.35 ± 3.67	21.29 ± 1.34	24.69 ± 0.94
Fibronectin 1 [µg/cm²]	25.02 ± 0.93	17.2 ± 0.34	17.56 ± 2.28	49.3 ± 4.28	20.54 ± 1.34	30.2 ± 5.6
Collagen 10 [µg/cm²]	17.08 ± 1.76	22.48 ± 0.66	22.36 ± 1.23	48.47 ± 3.44	18.37 ± 0.28	35.01 ± 4.09
Collagen 8 [µg/cm²]	19.67 ± 0.79	24.48 ± 2.18	21.93 ± 2.39	59.46 ± 3.97	21.75 ± 0.99	34.38 ± 0.93
Collagen 6 [µg/cm²]	16.83 ± 1.12	18.72 ± 1.11	18.91 ± 0.89	66.18 ± 1.29	22.96 ± 1.74	32.44 ± 2.69
Vitronectin 0.1 [µg/cm²]	28.67 ± 0.92	14.11 ± 1.54	16.81 ± 1.76	61.18 ± 3.97	21.47 ± 1.8	41.52 ± 5.57

Table 10: Adhesion time A_T [min] of human fibroblasts, GM-7373 endothelial cells, SH-SY5Y neuroblastoma cells, HaCaT keratinocytes, MG-63 osteoblasts, and A10 smooth muscle cells on adhesion ligands in dependence of the ligand concentration.
The results were given as average ± SEM of four independent measurements.

With respect to the adhesion time A_T [min] values, which were achieved on the maximum ligand concentrations, a cell specific ranking of adhesion ligands could be attested (Table 10). For fibroblasts the ranking was fibronectin (15.89 ± 1.34) > collagen (17.08 ± 1.76) > laminin (24.23 ± 2.67) > vitronectin (28.67 ± 0.92), for endothelial cells vitronectin (14.11 ± 1.54) > laminin (17.03 ± 1.28) > collagen (22.48 ± 0.66) > fibronectin (17.2 ± 0.34), for neuroblastoma cells vitronectin (16.81 ± 1.76) > laminin (16.89 ± 1.99) > fibronectin (17.57 ± 0.71) > collagen (22.36 ± 1.23), for keratino-cytes fibronectin (44.99 ± 3.17) > collagen (48.47 ± 3.44) > laminin (55.02 ± 5.56) > vitronectin (61.18 ± 3.97), for osteoblasts collagen (18.37 ± 0.28) > laminin ≅ fibronectin (about 20) > vitronectin (21.47 ± 1.8) and for smooth muscle cells laminin ≅ fibronectin (about 25) > collagen (35.01 ± 4.09) > vitronectin (41.52 ± 5.57). A reduced concentration of the preferred ligand correlated with an increase of adhesion time. Simultaneously, a reduced concentration of the not preferred ligand correlated with a decrease of adhesion time for all cell types except for neuroblastoma cells (Table 10).

Adhesion ligands influenced adhesion pattern in a cell specific manner

Adhesion pattern were analyzed on the maximun ligand coating concentrations (images not shown). In comparison to the control surface (Figure 34 a), laminin, and collagen, fibroblasts formed more contacts on fibronectin and less on vitronectin. Concerning endothelial cells, the contacts on laminin and vitronectin were more pronounced than on the control surface (Figure 34 b). On fibronectin and collagen they did not adhere with the whole cell body. The adhesion pattern of neuroblastoma cells was comparable between the control surface, fibronectin, and collagen (Figure 34 c). Closer contacts of the whole cell body were built on vitronectin and especially laminin. For keratinocytes no differences between the treatments were found (Figure 34 d). Similarly to the control surface, osteoblasts formed focal contacts on laminin and fibronectin, but contacts with the whole cell body on vitronectin and collagen (Figure 34 e). On all adhesion ligands smooth muscle cells adhered with the whole cell body like on the control (Figure 34 f). Small differences occurred as the contacts on laminin seemed to be closer followed by fibronectin and collagen, and less on vitronectin.

To get more insights into the effects, the adhesion pattern were quantified via the total gray scale distribution and average of the histograms. Histograms are exemplarily shown in Figure 35.

		Control	Laminin 2 μg/cm²	Fibronectin 5 μg/cm²	Collagen 10 μg/cm²	Vitronectin 0.1 μg/cm²
Fibroblasts	Minimum	14	14	22	22	30
	Maximum	196	164	157	158	157
	Average ± SEM	67.71 ± 2.21	63.73 ± 1.8 **	64.13 ± 1.93 **	66.59 ± 1.61	69.94 ± 1.66 *
Endothelial	Minimum	15	23	20	27	25
	Maximum	164	160	155	162	162
	Average ± SEM	67.3 ± 2.26	64.69 ± 2.34	67.97 ± 1.84	65.93 ± 2.34	61.63 ± 2.21 ***
Neuro-blastoma	Minimum	41	29	17	21	27
	Maximum	157	135	151	153	138
	Average ± SEM	68.71 ± 1.46	66.18 ± 1.62 *	72.37 ± 2.21 *	72.46 ± 2.16 *	70.63 ± 2.11
Keratino-cytes	Minimum	33	34	32	33	36
	Maximum	165	167	148	161	158
	Average ± SEM	69.97 ± 2.06	71.46 ± 2.31	68.82 ± 2.44	69.02 ± 2.28	74.24 ± 2.05 **
Osteoblasts	Minimum	34	40	39	36	26
	Maximum	121	158	136	132	148
	Average ± SEM	70.82 ± 1.84	75.24 ± 1.74 **	73.27 ± 1.83	74.89 ± 1.68 **	75.17 ± 2.39 *
Smooth muscle	Minimum	43	44	48	43	40
	Maximum	151	139	136	167	170
	Average ± SEM	86.64 ± 2.93	87.71 ± 2.96	88.21 ± 2.86	85.59 ± 2.8	86.71 ± 3.33

Table 11: Quantification of adhesion pattern of human fibroblasts, GM-7373 endothelial cells, SH-SY5Y neuroblastoma cells, HaCaT keratinocytes, MG-63 osteoblasts, and A10 smooth muscle cells in dependence of adhesion ligands after 24 h cultivation time.
The results were given as minimum, maximum and average ± SEM of gray values within each cell body of at least 100 cells per treatment.
* statistical difference ($p < 0.05$) in comparison to the control via Students-t-test analysis.

The control gray value distribution of fibroblasts was ranged in values of 14 to 196. On laminin the minimum value was reproduced, whereas on the other ligands it was increased to 22 and even 30 on vitronectin (Table 11). The maximum values were all decreased to about 160. On laminin and fibronectin fibroblasts decreased the average gray value significantly to about 64, whereas the value for collagen was comparable with the control with about 67. On vitronectin it was significantly increased to 70 (Table 11).

In contrast to fibroblasts, the maximum gray values of endothelial cells did not differ and were ranged between 155 and 164 for all treatments (Table 11). The minimum value of 15 on the control was increased on all ligands, especially on collagen to 27. On laminin, fibronectin, and collagen the average distance to the surface was comparable with the control surface of about 67. It was significantly decreased to 61 when cultivated on vitronectin (Table 11).

Neuroblastoma cells had a gray distribution of 41 to 157 and an average value of 69 under control conditions (Table 11). The average gray value was significantly switched to lager cell-surface distances of about 72 on fibronectin and collagen. On laminin it was significantly reduced to 66. The minimum values were decreased on all ligands to 17 - 29. On fibronectin and collagen the maximum value was comparable with the control surface of about 157, but it was reduced to about 137 on laminin and vitronectin (Table 11).

No differences in minimum and maximum gray values were observed for keratinocytes on all treatments (Table 11). They were in the same range of 31 to 165. The average gray value of 70 on the control surface was comparable with the average gray values on laminin, fibronectin, and collagen. On vitronectin it was significantly increased to 74 (Table 11).

On the control surface and on fibronectin an average gray value of about 71 was found for osteoblasts. On the other ligand surfaces the average gray values were switched to about 75 (Table 11). Furthermore, the gray value distribution of 31 to 121 on the control surface was changed on all ligands: on laminin to 40 – 158, on fibronectin to 39 – 136, on collagen to 36 – 132, and on vitronectin to 26 - 148 (Table 11).

The average cell to surface distances of smooth muscle cells were comparable on all treatments with 85 - 88. This similarity was also found for the minimum gray value of 40 - 48. The maximum value of 151 on the control was decreased to about 138 on laminin and fibronectin. On collagen and vitronectin it was increased to about 170 (Table 11).

Adhesion ligands affected cell morphology in a cell specific manner

Ligands effects on cell morphology were quantified by nucleus and cell dilation given as L_n/W_n and L_c/W_c. Furthermore, the average number of extensions was determined.

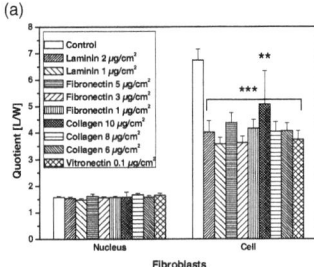
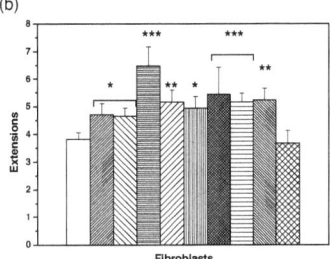

Figure 44: Quantification of cell morphology of human fibroblasts cultivated on adhesion ligands in comparison to the control over 24 h cultivation time.
(a) Quotient [L/W] of nucleus and cell, (b) number of extensions. The results were given as average ± SEM of at least 100 cells per treatment.
*** statistical difference (p < 0.05; 0.01; 0.001) in comparison to the control via Students-t-test analysis.

The nucleus dilation given as L_N/W_n of fibroblasts was not significantly changed by the ligands and comparable values of 1.59 like on the control surface were found (Figure 44 a). Under control conditions fibroblasts had a cell dilation of 6.74 ± 0.41. L_c/W_c was significantly reduced on all ligands to values of about 5 to 3.5 (Figure 44 a). Furthermore, the average number of extensions was determined. On vitronectin and on the control surface 3.67 ± 0.45 and 3.82 ± 0.24 extensions were formed, respectively (Figure 44 b). On fibronectin (5 µg/cm²) and collagen (10 µg/cm²) an increase was found to 6.47 ± 0.69 and 5.44 ± 0.49, respectively. A decreasing concentration of these ligands correlated with a decreasing number of extensions to values of about 5. On both laminin treatments the cells formed about 4.7 extensions (Figure 44 b).

Similarly to fibroblasts, the nucleus dilations of endothelial cells were not affected by the presence of adhesion ligands and reached comparable values like under control conditions of 1.39 ± 0.03 (Figure 45 a). The cell dilation given as L_c/W_c of 4.68 ± 0.99 on the control surface was significantly decreased on all ligands. The largest decrease was achieved on fibronectin (5 µg/cm²) with 1.84 ± 0.09 (Figure 45 a). A reduced concentration of fibronectin correlated with an increase of L_c/W_c. On the control 3.27 ± 0.1 extensions were formed (Figure 45 b). On both laminin treatments the cells formed significantly more extensions with comparable values of 4.4. The highest amount was reached when cultivated on vitronectin with 4.7 ± 0.38. On fibronectin (5 µg/cm²) the number of extensions was reduced significantly to 1.62 ± 0.23, followed by collagen (10 µg/cm²) 2.44 ± 0.28. On the other fibronectin and collagen treatments the number of extensions was increased. On 3 µg/cm² fibronectin and on 8 µg/cm² collagen the number of extensions was comparable with the control surface. On 1 µg/cm² fibronectin and 6 µg/cm² collagen significantly more extensions were formed than under control conditions (Figure 45 b).

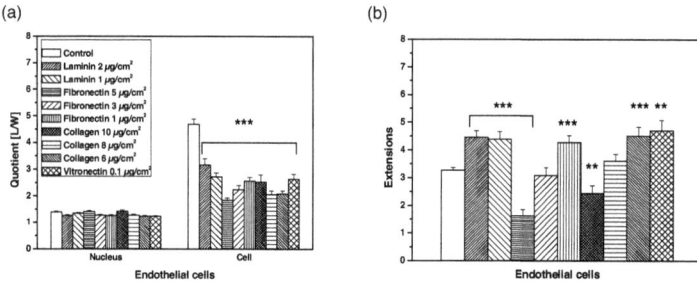

Figure 45: Quantification of cell morphology of GM-7373 endothelial cells cultivated on adhesion ligands in comparison to the control over 24 h cultivation time.
(a) Quotient [L/W] of nucleus and cell, (b) number of extensions. The results were given as average ± SEM of at least 100 cells per treatment.
*** statistical difference ($p < 0.05$; 0.01; 0.001) in comparison to the control via Students-t-test analysis.

For neuroblastoma cells a nucleus dilation of 2.35 ± 0.09 on the control surface was found. L_n/W_n was reduced on all ligands to about 1.6 (Figure 46 a). The cell dilation of 5.77 ± 0.39 on the control surface was reduced to 4.32 ± 0.3 on vitronectin and reduced to about 3.5 on laminin (Figure 46 a). The largest decrease was found for fibronectin (5 µg/cm²) to 1.95 ± 0.14. L_c/W_c on fibronectin was dependent on its coating-concentration. This concentration dependency was also observed for collagen with the lowest value of 2.47 ± 0.18 at a concentration of 10 µg/cm² (Figure 46 a). The quantification of extensions revealed that on vitronectin and laminin (2 µg/cm²) the cells formed as many extensions as on the control surface of about 5. On the other adhesion ligands, the number was reduced significantly. The largest reduction of extensions was found on fibronectin (5 µg/cm²) with 1.5 ± 0.15 (Figure 46 b).

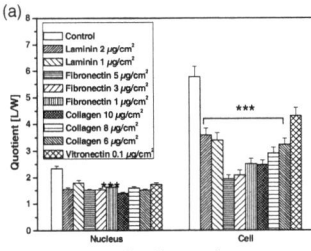 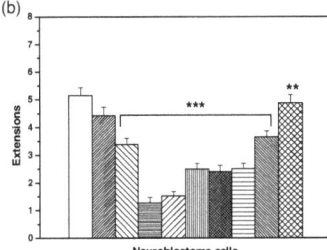

Figure 46: Quantification of cell morphology of SH-SY5Y neuroblastoma cells cultivated on adhesion ligands in comparison to the control over 24 h cultivation time.
(a) Quotient [L/W] of nucleus and cell, (b) number of extensions. The results were given as average ± SEM of at least 100 cells per treatment.
*** statistical difference (p < 0.05; 0.01; 0.001) in comparison to the control via Students-t-test analysis.

For keratinocytes the nucleus and cell dilations on the ligands were all comparable with the control surface of 1.44 ± 0.03 and 2.25 ± 0.13, respectively (Figure 47 a). Concerning the number of extensions, on both laminin coatings, vitronectin, fibronectin (1 μg/cm²), and collagen (6 μg/cm²) the number was comparable with the control of 2.62 ± 0.23 (Figure 47 b). The number was significantly increased on fibronectin (3 and 5 μg/cm²) and collagen (8 and 10 μg/cm²). The highest number of extensions was found on fibronectin (5 μg/cm²) and collagen (10 μg/cm²) with about 4.6 extensions per cell (Figure 47 b).

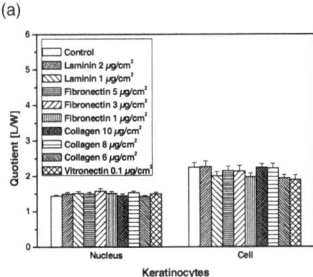 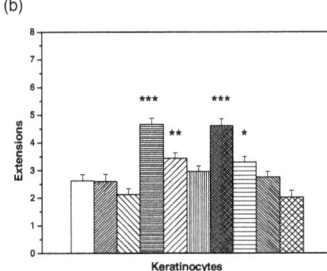

Figure 47: Quantification of cell morphology of HaCaT keratinocytes cultivated on adhesion ligands in comparison to the control over 24 h cultivation time.
(a) Quotient [L/W] of nucleus and cell, (b) number of extensions. The results were given as average ± SEM of at least 100 cells per treatment.
*** statistical difference (p < 0.05; 0.01; 0.001) in comparison to the control via Students-t-test analysis.

For osteoblasts a nuclues dilation of 1.49 ± 0.04 was found. On fibronectin (1 μg/cm²) L_n/W_n was significantly increased to 1.65 ± 0.1. It was not changed by the other ligands (Figure 48 a). The cell dilation of the control surface with 4.8 ± 0.32 was comparable on vitronectin and fibronectin. L_c/W_c was significantly reduced to 3.56 ± 0.25 on laminin (1 μg/cm²) and to 3.68 ± 0.28 on collagen (6 μg/cm²). The effects of fibronectin were not dependent on the coating concentration (Figure 48 a). Under control conditions and on 1 μg/cm² laminin about 3.21 extensions were formed (Figure 48 b). The number of extensions was significantly increased to 3.73 ± 0.17 on 2 μg/cm² laminin. The same concentration dependency was observed for collagen. On 10 μg/cm² collagen 5.24 ± 0.3 extensions were formed. The number of extensions decreased on lower collagen concentrations. On 8 μg/cm² collagen the found number was still significant higher than on the control surface (Figure 48 b). For fibronectin the results were opposite. The largest number of extensions was found on 1 μg/cm² fibronectin with about 4.6. The highest amount of average extensions was reached when cultivated on vitronectin with 5.86 ± 0.32 (Figure 48 b).

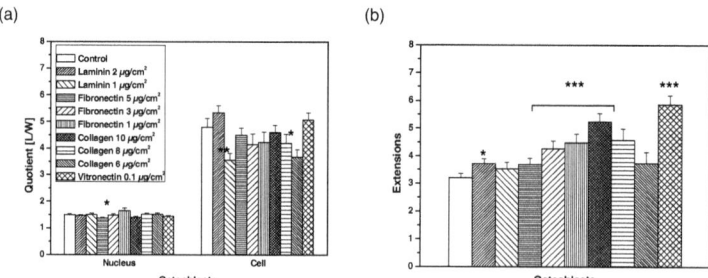

Figure 48: Quantification of cell morphology of MG-63 osteoblasts cultivated on adhesion ligands in comparison to the control over 24 h cultivation time.
(a) Quotient [L/W] of nucleus and cell, (b) number of extensions. The results were given as average ± SEM of at least 100 cells per treatment.
*** statistical difference ($p < 0.05$; 0.01; 0.001) in comparison to the control via Students-t-test analysis.

Smooth muscle cells presented a nucleus dilation of 1.53 ± 0.04 under control conditions. L_n/W_n was similar on all adhesion ligands (Figure 49 a). On the control surface, on laminin (2 μg/cm²) and fibronectin (5 μg/cm²) the cell dilations were comparable and ranged in 2.17 ± 0.11. L_c/W_c was significantly increased on collagen (10 μg/cm²) to 2.5 ± 0.11 and vitronectin to 2.66 ± 0.09. On 8 μg/cm² collagen L_c/W_c was comparable with the control surface, but it was significantly reduced on 6 μg/cm² collagen to 1.75 ± 0.09 (Figure 49 a). 4.47 ± 0.22 extensions were formed under control conditions. A significant increase in the number of extensions was found for 2 μg/cm² laminin and 10 μg/cm² collagen to 6.37 ± 0.38 and 6.22 ± 0.28, respectively (Figure 49 b). On 1 μg/cm² laminin the number of extensions was comparable with the control surface. On 8 μg/cm² and the number of extensions was reduced to about 3.76 ± 0.19. This reduction was significant when cultivated on 6 μg/cm² collagen. Vitronectin and fibronectin did not affect the formation of extensions (Figure 49 b).

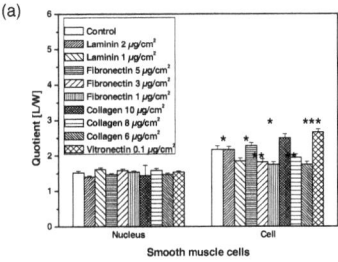 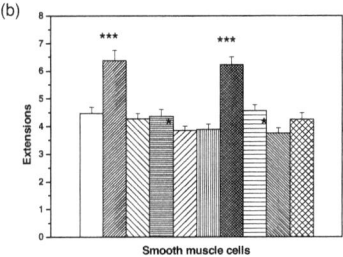

Figure 49: Quantification of cell morphology of A10 smooth muscle cells cultivated on adhesion ligands in comparison to the control over 24 h cultivation time.
(a) Quotient [L/W] of nucleus and cell, (b) number of extensions. The results were given as average ± SEM of at least 100 cells per treatment.
*** statistical difference (p < 0.05; 0.01; 0.001) in comparison to the control via Students-t-test analysis.

Adhesion ligands affected proliferation in a cell specific manner

The proliferation profiles were normalized on the starting cell densities [cells/ml] of $1.14*10^4$ for human fibroblasts, $6.16*10^4$ for GM-7373 endothelial cells, $3.34*10^4$ for SH-SY5Y neuroblastoma cells, $2*10^5$ for HaCaT keratinocytes, $6.37*10^4$ for MG-63 osteoblasts, and $2.06*10^4$ for A10 smooth muscle cells.

After 48 h cultivation time fibroblasts proliferated to 271.72 % ± 28.68 on the control surface (Figure 50 a). On fibronectin (5 μg/cm²) a cell density [%] of 317.69 ± 29.68 was found. Whereas a decreased concentration of fibronectin correlated with a reduction of proliferation even lower than on the control, no concentration dependency was observed for collagen. Cultivated on laminin and vitronectin the proliferation was reduced to values of about 220 % in comparison to the control. The growth behavior on laminin with 1 μg/cm² was better than on 2 μg/cm² (Figure 50 a).

In contrast to fibroblasts, endothelial cells proliferated comparably on the control surface, vitronectin, and laminin (2 µg/cm²) with a cell density of about 240 % after 48 h cultivation time (Figure 50 b). The effect of laminin was dependent on the concentration, since the proliferation was reduced to 209.61 % ± 18.3 on 1 µg/cm². The cells proliferated even less when cultivated on fibronectin (5 µg/cm²) with 189.03 % ± 23.39 and on collagen (10 µg/cm²) with 203.22 % ± 30.96. A decrease of fibronectin and collagen concentration correlated with an increase of cell densities up to 220 % (Figure 50 b).

Under control conditions a cell density of 392.24 % ± 34.6 was found for neuroblastoma cells (Figure 50 c). On laminin (2 µg/cm²) and vitronectin the cells proliferated to 436.11 % ± 32.23 and to 435.61 % ± 33.56, respectively. On laminin (1 µg/cm²) a proliferation of 450.02 % ± 17.49 was observed. Concerning fibronectin and collagen cell growth was reduced on the maximum concentrations to 331.49 % ± 35.76 and 390.23 % ± 3.94, respectively. It was increased on lower concentrations of fibronectin and collagen to comparable cell densities like on the control surface (Figure 50 c).

The observed ligand effects on keratinocytes (Figure 50 d) and osteoblasts (Figure 50 e) were comparable. After 96 h cultivation time the cells proliferated on the maximum ligand concentrations to the same cell densities [%] as under control conditions of 611.44 ± 32.33 and 705.68 ± 6.43, respectively. A reduction of ligand concentration always caused a decrease of growth behavior. For both cell types the lowest cell densities were found on fibronectin (1 and 3 µg/cm²) of about 530 % and 520 %, respectively (Figure 50 d, e).

Results

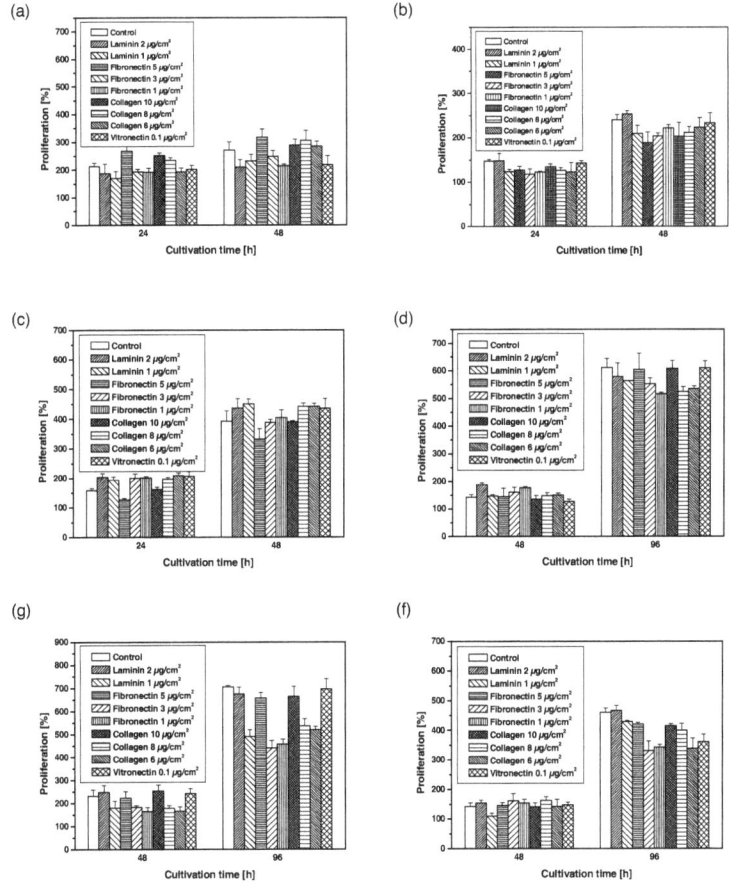

Figure 50: Proliferation profiles of (a) human fibroblasts, (b) GM-7373 endothelial cells, (c) SH-SY5Y neuroblastoma cells, (d) HaCaT keratinocytes, (e) MG-63 osteoblasts, and (f) A10 smooth muscle cells in dependence of adhesion ligands in comparison to the control.
The results were given as average ± SEM of four independent measurements, normalized in percent on the starting cell density at time t = 0 h ($1.14*10^4$, $6.16*10^4$, $3.34*10^4$, $2*10^5$, $6.37*10^4$, and $2.09*10^4$ cells/ml, respectively). The cultivation times varied between 48 and 96 h.

Smooth muscle cells proliferated to 460.02 % ± 14.92 under control conditions. On laminin (2 µg/cm²) a cell density [%] of 467.4 ± 16.25, on fibronectin (5 µg/cm²) of 421.04 ± 5.37, on collagen (10 µg/cm²) of 414.75 ± 6.77, and on vitronectin of 361.67 ± 25.09 were found. A decreased ligand concentration reduced the proliferation (Figure 50 f).

To get more insights into the ligand effects on proliferation the doubling time [h] was calculated for each treatment.

Cell types	Doubling time [h] ± SEM					
	Fibro-blasts	Endo-thelial	Neuro-blastoma	Keratino-cytes	Osteo-blasts	Smooth muscle
Control	27.39 ± 1.01	39.22 ± 2.19	26.88 ± 2.39	46.51 ± 3.22	33.53 ± 3.22	50.73 ± 0.51
Laminin 2 [µg/cm²]	40.35 ± 4.64 *	39.33 ± 6.06	23.06 ± 1.72	47.52 ± 3.01	38.63 ± 4.83	49.18 ± 1.36
Laminin 1 [µg/cm²]	38.82 ± 2.99 *	50.24 ± 4.08	23.13 ± 1.04	45.88 ± 1.64	46.5 ± 4.76	61.6 ± 2.33 ***
Fibronectin 5 [µg/cm²]	24.57 ± 0.94	54.54 ± 8.66	32.56 ± 4.51	42.16 ± 4.6	40.87 ± 1.58	54.96 ± 1.87
Fibronectin 3 [µg/cm²]	33.81 ± 2.61	42.66 ± 1.6 ***	24.4 ± 0.75	46.58 ± 2.56	48.11 ± 0.89 *	56.3 ± 2.17
Fibronectin 1 [µg/cm²]	40.46 ± 2.06 **	48.86 ± 2.42 *	23.96 ± 0.62	47.67 ± 2.24	53.83 ± 3.43 *	56.64 ± 2.72
Collagen 10 [µg/cm²]	25.27 ± 0.95	57.2 ± 10.43	28.81 ± 0.62	44.89 ± 2.09	37.17 ± 3.91	54.63 ± 2.78
Collagen 8 [µg/cm²]	24.58 ± 3.81	48.62 ± 3.72 *	22.85 ± 0.25	47.57 ± 2.06	44.9 ± 1.77	57.57 ± 4.77
Collagen 6 [µg/cm²]	32.8 ± 2.18	55.15 ± 8.93	22.47 ± 0.27	46.46 ± 0.56	48.47 ± 2.09 *	57.44 ± 1.66 **
Vitronectin 0.1 [µg/cm²]	44.86 ± 4.22 *	42.25 ± 3.55	23.08 ± 1.46	50.35 ± 3.22	34.06 ± 2.14	56.79 ± 1.76 ***

Table 12: Doubling time [h] of human fibroblasts, GM-7373 endothelial cells, SH-SY5Y neuroblastoma cells, HaCaT keratinocytes, MG-63 osteoblasts, and A10 smooth muscle cells on adhesion ligands in dependence of the ligand concentration.
The results were given as average ± SEM of four independent measurements.
*** statistical difference (p < 0.05; 0.01; 0.001) in comparison to the control via Student's-t-test analysis.

A doubling time of 27.39 h ± 1.01 was found for fibroblasts under control conditions. This time was comparable on fibronectin (5 and 3 µg/cm²) and on collagen (10, 8 and 6 µg/cm²). On 1 µg/cm² fibronectin, on both laminin treatments and vitronectin the doubling time was significantly increased to more than 39 h (Table 12).

Endothelial cells had similar doubling times the control surface with 39.21 h ± 2.19, on laminin (2 µg/cm²) and on vitronectin. On laminin (1 µg/cm²), fibronectin, and collagen the found doubling times were larger than 48 h. A reduced concentration of fibronectin and collagen reduced the doubling time to about 48 h (Table 12).

Under control conditions a doubling time [h] of 26.88 ± 2.39 occured for neuroblastoma cells. On laminin (2 and 1 µg/cm²), vitronectin, fibronectin (3 and 1 µg/cm²) and collagen

(8 and 6 µg/cm²) the doubling time was reduced to about 23 h when compared with the control surface (Table 12). On fibronectin (5 µg/cm²) and collagen (10 µg/cm²) the doubling time [h] was increased to 32.56 ± 4.51 and 28.81 ± 0.62, respectively (Table 12).

Keratinocytes showed a doubling time of 46.51 h ± 0.64 on the control surface. On fibronectin (5 µg/cm²) and collagen (10 µg/cm²) this time was reduced to about 42 h. On laminin (2 µg/cm²) and vitronectin it was increased to about 50 h. On the other treatments the cells grew as fast as under control conditions (Table 12).

On the control surface a doubling time of 33.53 h ± 3.22 was found for osteoblasts This time [h] was increased on vitronectin to 34.06 ± 2.14, followed by collagen (10 µg/cm²) to 37.17 ± 3.91, laminin (2 µg/cm²) to 38.63 ± 4.83, and fibronectin (5 µg/cm²) to 40.87 ± 1.58 (Table 12). On fibronectin (3 and 1 µg/cm²) and collagen (6 µg/cm²) the increase of the doubling time was significant (Table 12).

Smooth muscle cells presented a doubling time [h] of 50.73 ± 0.51 under control conditions. On laminin (2 µg/cm²) the doubling time was comparable. An significant increase in the doubling time [h] was found on laminin (1 µg/cm²) to 61.6 ± 2.33 and on collagen (6 µg/cm²) to 57.44 ± 1.66 (Table 12). On the other ligands it was reduced in the order collagen ≅ fibronectin followed by vitronectin to 56.79 h ± 1.76. The effect of vitronectin was significant (Table 12).

Adhesion ligands affected gap junction coupling in a cell specific manner

Gap junction coupling was analyzed by scrape loading method. Exemplarily, the images of GM-7373 endothelial cells are shown in Figure 51. The results were quantified as the diffusion distance of lucifer yellow.

Under control conditions fibroblasts showed an average diffusion distance of 160.31 px ± 19.52. This distance was increased when cultivated on fibronectin and collagen. This effect did not correlate with the order of ligand concentration (Table 13). The largest distances were found for fibronectin (3 µg/cm²) with 200.1 px ± 2.77 and collagen (8 µg/cm²) with 186.18 px ± 13.61. On 1 µg/cm² laminin the diffusion distance of 149.92 px ± 17.15 was larger than on 2 µg/cm² laminin, but in both cases shorter than on the control. On vitronectin gap junction coupling was significantly reduced to 91.19 px ± 5.79 (Table 13).

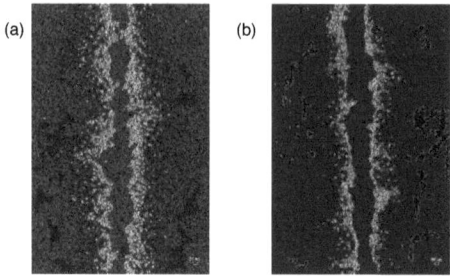

Figure 51: Fluorescence images of cell to cell communication over gap junction channels visualized with lucifer yellow after scrape loading procedure.
Performed with GM-7373 endothelial cells after 24 h cultivation time (a) long diffusion distance, (b) short diffusion distance.

In comparison to the control surface with a found diffusion distance [px] of 131.12 ± 7.18, gap junction coupling of endothelial cells was increased by the presence of vitronectin to 142.84 ± 4.03 (Table 13). On all the other ligands the diffusion distances were smaller than on the control. The influences of fibronectin and collagen were not dependent on the concentration. Comparable values in a range of 105 - 110 px were observed. Varying laminin concentrations affected the coupling selectively. The diffusion distance was larger on 2 μg/cm² laminin with 120.34 px ± 3.15 than on 1 μg/cm² laminin with 104.47 px ± 4.88. The observed effects of collagen (8 μg/cm²), laminin and fibronectin (both with 1 μg/cm²) were significant (Table 13).

Keratinocytes showed a diffusion distance [px] of 116.96 ± 4.6 on the control surface. On collagen (10 and 6 μg/cm²) the diffusion distance [px] was reduced to 105.21 ± 5.19 and 109.22 ± 8.22 (Table 13). It was increased to 138.33 ± 21.13 on 8 μg/cm² collagen. A significant increase to 140.18 px ± 6.97 of the diffusion distance was found on 3 μg/cm² fibronectin. The distance was reduced on the other fibronectin concentrations, but not in a concentration dependent manner. The comparable values of about 120 px for both laminin concentrations were higher than on the control. On vitronectin the diffusion distance [px] was reduced to 107.66 ± 1.88 (Table 13).

The diffusion distance on the control with 112.35 px ± 8.79 of smooth muscle cells was reduced on 2 and 1 µg/cm² laminin to 90.55 px ± 10.09 and 93.76 px ± 15.06, respectively (Table 13). Furthermore, a decreased diffusion distance was also found for vitronectin with 86.47 px ± 6.53. Independent of the collagen concentration, distances of about 115 px were reached, which were similar to the control (Table 13). An increase in the diffusion distance to 129.35 px ± 13.92 was observed on 5 µg/cm² fibronectin. The shortest diffusion distance occured on 1 µg/cm² fibronectin with 85.91 px ± 8.85.

Cell type	Diffusion distance [px] of lucifer yellow ± SEM			
	Fibroblasts	Endothelial cells	Keratinocytes	Smooth muscle
Control	160.31 ± 19.52	131.12 ± 7.18	116.96 ± 4.6	112.35 ± 8.79
Laminin 2 [µg/cm²]	121.61 ± 6.22	120.34 ± 3.15	125.69 ± 3.11	90.55 ± 10.09
Laminin 1 [µg/cm²]	149.92 ± 17.15	104.47 ± 4.88 *	120.19 ± 2.57	93.76 ± 15.06
Fibronectin 5 [µg/cm²]	184.49 ± 18.76	108.89 ± 10.51	135.9 ± 14.1	129.35 ± 13.92
Fibronectin 3 [µg/cm²]	200.1 ± 2.77	109.59 ± 9.38	140.18 ± 6.97 *	104.2 ± 8.95
Fibronectin 1 [µg/cm²]	130.89 ± 16.41	102.75 ± 4.92 *	121.57 ± 5.78	85.91 ± 8.85
Collagen 10 [µg/cm²]	126.23 ± 11.37	112.18 ± 8.27	105.21 ± 5.19	113.03 ± 13.14
Collagen 8 [µg/cm²]	186.18 ± 13.61	105.61 ± 7.04	138.33 ± 21.13	115.4 ± 6.03
Collagen 6 [µg/cm²]	174.19 ± 15.65	105.8 ± 8.78	109.22 ± 8.06	117.12 ± 10.53
Vitronectin 0.1 [µg/cm²]	91.19 ± 5.79 *	142.84 ± 4.03	107.66 ± 1.88	86.47 ± 6.53

Table 13: Quantification of gap junction coupling of human fibroblasts, GM-7373 endothelial cells, HaCaT keratinocytes, and A10 smooth muscle cells on adhesion ligands in dependence of the ligand concentration.
Gap junction coupling was examined with the help of the scrape loading procedure with lucifer yellow in comparison to the control after 24 h cultivation time. The results were given as average of the diffusion distance [px] ± SEM of 16 images per treatment coming from four independent measurements.
* statistical difference ($p < 0.05$) in comparison to the control via Student's-t-test analysis.

6 Discussion

In the field of tissue engineering and biomedical research the design and selection of biomaterials is of huge importance, since the interaction with cells promote the desired biological functions [5]. For this purpose, all material variables influencing cell functions and tissue morphogenesis have to be taken into account.

In this work, a wide range of materials with varying properties, tissue-engineered constructs, cell transfer, and topographically-functionalized biomaterials were tested for their biomedical application with focus on cellular responses. Furthermore, the influence of diverse adhesion ligands on cells were examined. Cell behavior was characterized via DNA damage effects, adhesion, morphology, orientation, and proliferation. All measurements were performed with different cell types to explore, whether material effects occurred in a cell specific manner.

6.1 Cell responses to three-dimensional scaffolds

One promising approach in tissue engineering relies on the application of three-dimensional scaffolds that serve as substitutes for tissues and organs to be replaced or support body's own regeneration [13]. For this purpose, scaffolds have to fulfill several requirements related to positive tissue interaction, fluid and nutrient exchange, and vascularization. From the technical point of view, scaffolds have to consist of materials with appropiate mechanical properties. Moreover, ones need a technique that enables the fabrication of any tissue-engineered constructs.

For this work, scaffolds were produced with the two-photon polymerization technique. It was carried out by Dr. A. Ovsianikov at the Laser Zentrum Hannover e. V. (Germany). This method has several advantages over others due to the design of any desired three-dimensional structure down to a resolution of 100 nm, the active control of structure dimensions and features, cost-effectiveness, and reproducibility [22, 23]. For this purpose, the potential material candidates Ormocomp® and PEG were tested for their application. Parallel to the development of scaffolds, unstructured Ormocomp® and PEG with various properties were used to estimate cellular responses in general.

6.1.1 Ormocomp® does not negatively affect cellular behavior

The comet assay revealed that the presence of Ormocomp® did not increase the incidence of DNA damage effects of GFSHR-17 granulosa cells (page 39). Furthermore, it was demonstrated that granulosa cells, GM-7373 endothelial cells, and SH-SY5Y neuroblastoma cells proliferated at the same rate on Ormocomp® like under control conditions (Figure 13 a - c). This observation was also supported by the parameter doupling time. Since the doubling times found on Ormocomp® were comparable with the control treatment, it can be concluded that the polymer did not negatively influenced cell cycle progression (Figure 13 d). Ormocomp® was also shown to be compatible with the formation of cellular junctions. With the help of the patch-clamp technique no significant changes in gap junction conductions were found for granulosa cells cultivated on Ormocomp® [22]. These results demonstrate, that Ormocomp® did not negatively affect cellular behavior cell type independently. Therefore, it is a promising material for biomedical applications.

6.1.2 The biomedical use of PEG depends on its composition and cell type

For designing suitable biomaterials several parameters such as molecular weight, the used photoinitiator and material aging have to be considered related to their influence on cellular behavior. These factors are very important for the fabrication of three-dimensional scaffolds, as they correlate with the structural resolution and the size dimensions that shall be produced with the help of laser technologies [22]. A good candidate for analyzing these effects on different cell types is hydrogel PEG.

First, DNA damage effects of two PEG compositions (SR259 and SR610, both supplemented with 2 wt% photoinitiator Bis) that differed in their molecular weight were examined by comet assay. It was found that PEG SR259 significantly increased the incidence of DNA strand breaks of GFSHR-17 granulosa cells (page 39). As both samples were produced with the same photoinitiator and concentration it can be assumed that the investigated DNA damages correlated with the molecular weight. Further analysis are needed to clarify this topic. Since PEG samples with photoinitiator Bis were not stable under in vitro conditions, in the following PEG was photo-crosslinked with 2 wt% photoinitiator Irgacure 2959. The poor material stability of PEG supplemented with Bis remains unclear.

Freshly prepared PEG SR610 samples including 2 wt% Irgacure 2959 also caused DNA damage effects cell type independently. The tailmoments of human fibroblasts, GM-7373 endothelial cells, and SH-SY5Y neuroblastoma cells were significantly increased in the presence of PEG in comparison to the control (Table 1). Furthermore, proliferation was clearly reduced for all cell types (Figure 16). It is suggested that the observed DNA damage effects and the decrease of cell growth were caused by possible residuals, such as photoinitiators or monomers, which were not consumed during the photo-polymerization reaction of PEG. For this reason material aging was suggested to overcome the toxicity of Irgacure 2959. By this procedure the material is placed in water for seven days. The influences on DNA strand breaking and proliferation were analyzed in dependence of PEG aging. After material aging no significant changes in the tailmoments of all cell types such as fibroblasts, endothelial, and neuroblastoma cells were observed in comparison to the control (Table 1). Probably, material aging has removed the toxic residuals of PEG. DNA damages identified as irreparable double strand breaks are crucial criteria for apoptosis and necrosis, which can all be examined by comet assay [109].

Concerning cell growth a selective cell control was found on aged PEG SR610 (supplemented with 2 wt% photoinitiator Irgacure 2959). While fibroblasts proliferated at the same rate like under control conditions (Figure 16 a), endothelial and neuroblastoma cells reduced the cell densities over the total cultivation time (Figure 16 b, c).

For adherent cells like fibroblasts, endothelial cells, and neuroblastoma cells proliferation is only possible when they attach to material surface. As adhesion is considered to be the first step in biomaterial cell interaction that activates and guides cellular behavior like proliferation [59], it can be suggested that PEG influenced adhesion in a cell specific manner. With respect to the proliferation results (Figure 16), it can be proposed that PEG inhibits the adhesion of endothelial cells and neuroblastoma cells, but not the adhesion of fibroblasts. This thesis is supported by the analysis of adhesion time A_T. While fibroblasts adhered faster on PEG than on the control, endothelial cells and neuroblastoma cells adhered slower (Table 2). Furthermore, it can be supposed that the found selective cell control is caused by cell specific differences in adhesion mechanism. To explain this effect in detail, further analysis are needed that identify possible differences in the adhesion mechanism of the cells and selective material effects on adhesion. These facts point out, that the knowledge of cell specific differences in adhesion mechanism is a key consideration when developing materials and tissue-engineered constructs for biomedical applications.

On the one hand the found selective cell control of PEG has carefully be taken into account for tissue-engineered constructs. On the other hand this finding opens new possible applications for this material with respect to the improvement of implant adaptation.

6.1.3 Cell localization on three-dimensional scaffolds

To improve the functionality of tissue-engineered constructs, research has turned towards the creation of cell-coated implants that mimic the native tissue with respect to anatomical geometry and cell placement [16]. Coming to three-dimensional scaffolds, the questions are, whether the cells fall within the features, lay on the top or adhere on lateral surfaces, whether they present their normal morphology, and are able to proliferate.

With the help of two-photon polymerization technique three-dimensional grating structures composed of Ormocomp® (Figure 18) and rings composed of PEG (Figure 17) with varying diameters were produced. A microscopic study with human fibroblasts, NIH3T3 fibroblasts, and GM-7373 endothelial cells revealed, that all cell types fell into the structures with diameters larger than 30 μm (Figure 18 e, f, Figure 21, Figure 20). Attaching to the surface, the cells were able to proliferate over a longer period of cultivation time up to ten days indicating that the structure features provided nutrient exchange needed for cell cycle progression. On the grating scaffolds with 30 and 40 μm in diameter the cells adapted their morphology to feature dimensions, and also adhered on the top (Figure 18 b, c). These findings point out, that the localization of the cells was dependent on scaffold dimensions, and independent from the cell type and applied material. However, with this experimental procedure no cells attached on lateral surfaces related to the sedimentation of the cells.

For future applications of three-dimensional scaffolds the size dimensions have carefully taken into account. Furthermore, the mechanical properties of the materials are a critical factor. The microscopic documentation demonstrated, that scaffolds composed of PEG SR610 were not stable under in vitro conditions (Figure 21, Figure 20). This fact is related to the native properties of hydrogels – they swell in aqueous solution followed by the deviation of structure geometry from its original design. It was shown that hydrogel swelling correlates with its crosslinking density and molecular weight [11, 110]. On the one hand, this property is necessary for material degradation and the use of drug-delivery vehicles. On the other hand, a hydrogel composition with appropiate mechanical properties has to be choosen that realizes the fabrication of scaffolds and defined design endurance.

6.1.4 Cells adhere on lateral surfaces

Traditional cell seeding caused a sedimentation of cells and lateral surfaces were not coated with cells. To overcome a sedimentation of cells, they were kept in suspsension by the use of a shaking table (Figure 8). With this procedure it was possible, that SH-SY5Y neuroblastoma cells adhered on lateral surfaces of cyclindrical scaffolds composed of Ormocomp® (Figure 19). However, it led to a heterogeneous cell distribution. This phenomenon was caused by some issues related to the hydromechanics which were presented by the shaking table and automatically introduced into the cell culture system. To predict the formation of three-dimensional tissue substitutes, a combination of physical analysis, stimulation technology, and mechanical engineering is required [22].

6.2 Cell transport with laser-induced forward transfer

A more effective and controllable way to pre-coat three-dimensional scaffolds with cells, is the use of tissue and organ printing concepts realized by laser-printing technologies such as laser-induced forward transfer [16, 28, 106]. This technique was established by Dipl.-Ing. M. Grüne and Dr. L. Koch at the Laser Zentrum Hannover e. V. (Germany). Cells of interest for scaffold coating are autologous cells that reduce the risk of immune reaction, and stem cells which can differentiate in any cell type [15, 17].

Important preconditions for the successful use of laser-induced forward transfer rely on the controllable transport of cells and the defined cell arrangement on the target. From the biological point of view this transport shall not harm the cells with respect to DNA damage effects and proliferation. These parameters were estimated with NIH3T3 fibroblasts, HaCaT keratinocytes, and human and porcine mesenchymal stem cells.

It was found that the cells were transported and arranged in defined pattern with high precision. Furthermore, the possibility of creating specific pattern consisting of more than one cell type was demonstrated (Figure 22). Cell transport did not negatively affect the growth behavior of fibroblasts and keratinocytes when compared with the control conditions (Figure 23). The laser-induced forward transfer procedure did not significantly increase the incidence of DNA damage effects of the investigated cell types and stem cells (Table 3). Therefore, the laser-induced forward transfer offers new promising possibilities in the field of tissue engineering [106].

However, for the successful use of laser-induced forward transfer further analysis are needed. First, transport effects on stem cell differentiation have to be investigated. Moreover, multiple cell layers have to be produced, which for instance could be used for skin replacement. After that the cell behavior and integrety of the layers have to be examined. It also offers the possibiliy to study cell behavior in three-dimensions. The last step involves scaffold coating with cells. All analysis have to performed in dependence of the material placed on the target.

6.3 Selective control of cellular behavior in dependence of material chemistry

The knowledge of biomaterial cell interaction is a key consideration when developing implants with perfect tissue integration. With respect to foreign body reactions such as the formation of granulation tissue and fibrosis that minimize implant adaptation and function, research has turned forward to the finding of materials that provide a selective control of cellular behavior in dependence of its application [30, 40].

6.3.1 Selective control of cellular behavior in dependence of material hydrophobicity

On PEG a selective cell control of human fibroblasts, GM-7373 endothelial cells, and SH-SY5Y neuroblastoma cells was found with respect to adhesion time A_T (Table 2) and proliferation (Figure 16). Fibroblasts adhered faster on the material, while endothelial cells and neuroblastoma cells adhered slower. The cell specific influences on adhesion time were also reflected in the proliferation profiles. An increase of adhesion time correlated with a reduction of proliferation, as seen for endothelial cells

Discussion

and neuroblastoma cells (Figure 16 b, c). As adhesion guides other cellular responses such as proliferation, this finding supports that adhesion to material surface is the critical step in biomaterial cell interaction [59]. Since a selective cell control was found, it can be suggested that the adhesion mechanism are cell specific. However, the question is which property of the material is the critical key that enables a cell specifc influence on adhesion. As PEG is a hydrogel, it was suggested that the found selective cell control is caused by material hydrophilicity. For this purpose, the hydrogel HESHEMA and the hydrophobic silicone elastomer (Table 4) were used for further comparisons using human fibroblasts, GM-7373 endothelial cells, and SH-SY5Y neuroblastoma cells. Theoretically, if material hydrophobicity is the key for selective cell control, material effects of HESHEMA would be comparable with the effects of PEG, and simultaneously the effects of silicone elastomer would be opposite on the used cell types.

In comparison to the control endothelial cells and neuroblastoma cells adhered significantly slower on HESHEMA, whereas fibroblasts adhered faster (Table 2). The effects of HESHEMA on adhesion time were similar to the effects of PEG. On hydrophobic silicone elastomer this observation was opposite. Adhesion time of fibroblasts was increased, whereas endothelial cells and neuroblastoma cells accelerated the adhesion (Table 2). By this means the expected opposite effect of hydrophobic silicone elastomer was confirmed. The differences found for material influences on adhesion time were again reflected in the proliferation profiles. An increase of adhesion time correlated with a reduction of proliferation, a decrease of adhesion time with an increase in proliferation. Fibroblasts reduced the cell growth on hydrophobic silicone elastomer (Figure 14 a), while endothelial cells and neuroblastoma cells were not affected (Figure 14 b, c). This finding supports the important role of cellular adhesion in biomaterial cell interaction.

Obviously, cell types can be separated into two classes: cells preferring hydrophobic materials versus cells preferring hydrophilic materials. Positive responses of fibroblasts to hydrophilic and of endothelial cells to hydrophobic materials and the opposite were already described in literature [111 - 113]. In the present context it is important to ask, why material hydrophobicity caused a selective cell control. With respect to the order of cellular behavior, material hydrophobicity affects the adhesion of the cells [51, 52, 59]. However, a detailed explanation still remains unclear. Theoretically, the material could influence the association of the adhesion ligands coming from the extracellular matrix. It was found that the alignment, localization, concentration, and conformation of the ligands are governed by material properties [50]. After the association to material surface, the ligands bind specifically to the adhesion receptors integrins. In case the ligands do not offer a conformation on the material that enables the binding to the receptors or achieve a concentration on the material which is too small for cell binding, adhesion could be negatively affected or even inhibited. However, to answer this question, it needs to be known, which adhesion ligands are used by the specific cell types and whether the cells react to varying ligand concentrations. Since a selective cell control in dependence of material hydrophobicity was found, it can be supposed that the used ligands are cell specific. After identifying the used ligands it can be analyzed in what manner the material affects ligand association. These suggestions point out, that the knowledge of cell specificity of adhesion mechanism is a key consideration for biomedical research.

6.3.2 Cell control in dependence of material crosslinking density

In the past, hydrogels have gained widespread interest in biomedical applications due to controllable chemical and mechanical properties, the combination with bioactive molecules, degradation, and the fabrication of drug delivery vehicles [11, 15]. The findings in this work demonstrated further advantages of hydrogels. Material hydrophilicity was shown to be a promising parameter for selective cell control. Besides finding selective cell responses to hydrophilic HESHEMA, it was also examined, whether cell proliferation was dependent on the degree of substitution (DS). This parameter determines the crosslinking of hydrogels which correlates with material swelling, degradation, and stiffness. According to Bryant [114] a small DS value refers to a low crosslinking density causing a reduced material stiffness.

The growth profiles of human fibroblasts, GM-7373 endothelial cells, and SH-SY5Y neuroblastoma cells showed that on smaller DS values the proliferation was reduced cell type independently (Figure 15). Obviously, cells react to material crosslinking and stiffness. This observation was also reported by several studies [11, 115]. It was demonstrated that fibroblasts and endothelial cells prefer stiff materials [116]. The question how the information of an elastic material is converted into biochemical signaling responsible for cellular behavior was calculated in different mathematical models established by Nicolas [117]. These models refer to material influence on cellular adhesion. It was predicted that the formation, dynamics, and functions of focal adhesion complexes, which bind the cells to the surface via components of the extracellular matrix, integrin receptors, adaptor molecules, and the cytoskeleton, are dependent on material stiffness and thermodynamically limited by the elasticity and thickness of the extracellular matrix. These changes in mechanical forces can alter cytoskeletal structure and signal transduction. By this means the mechanosensitivity of integrins results in a mechanochemistry at the molecular level [118]. However, a detailed understanding of factors involved and influenced during these processes still remains unclear. But it can be pointed out that the knowledge of adhesion mechanism is essential for understanding cell control in dependence of material crosslinking and stiffness.

6.4 Selective cell control by surface topographies

The surface control of cellular behavior plays an important role in the formation of tissues and implant integration. For this purpose, different functionalization methods of materials have been established to achieve a selective cell control [15, 32, 33]. A promising approach is the topographical functionalization, which can be accomplished by laser processing via ablation. This measurement was established by Dipl.-Phys. E. Fadeeva at the Laser Zentrum Hannover e. V. (Germany). This method offers many advantages over others, namely low surface contamination, low mechanical damages, and controllable surface texturing with complicated geometries. The use of ultrashort pulsed lasers has additional benefits due to a better resolution and a reduced heat affected zone [41]. It was demonstrated that a large variety of structures such as simple roughness or defined surface topologies in micro- and nanometer scale can be produced in almost all solid materials. Moreover, topographical features and sizes were reproducible and controllable by establishing the same processing parameters (Figure 24, Figure 27, Figure 28) [42]. Laser-manufactured surface topographies of silicon can easiliy be transferred into soft materials such as silicone elastomer with the help of the negative replication process (Figure 25) [42]. This procedure was established by Dipl.-Phys. E. Fadeeva at the Laser Zentrum Hannover e. V. (Germany). Besides the wide range of possible materials and topologies, the generation of just one master sample by laser ablation accelerates the production time of such surface topographies.

Different surface characteristics such as groove and spike structures in micrometer scale, hierarchical nano- and micro- superimposed structures, nanoroughness- and grooves in different materials such as silicon, silicone elastomer, titanium, and platinum were produced by Dipl.-Phys. E. Fadeeva at the Laser Zentrum Hannover e. V. (Germany). In comparison to the control and unstructured samples no significant increase in DNA damages occurred for directly ablated topologies in silicon, titanium, and platinum and the negative replicas in silicone elastomer (Table 5). The parameter laser fluence did not cause any effects (Table 5). Cell type independently, the found tailmoments on the structures were comparable with the control treatment. Directly ablated features in silicone elastomer increased significantly DNA strand breaking of fibroblasts and neuroblastoma cells (Table 5). It can be assumed that laser radiation broke the long polymer chains forming free radicals which were responsible for the effect on the DNA strand breaks [43]. Further analysis are needed to clarifiy this topic. This result points out another advantage for the negative replication technique, as by this means silicone elastomer can still be used for a topographical functionalization.

All surface structures were used to analyze their effectiveness for selective cell control.

6.4.1 Selective control of orientation by groove structures

One important criteria addressed to implant specification, is the control of cell alignment and orientation. According to Tan [119] osteoblast orientation is essential for bone formation. Promising surface features enabling the improve and control of cell orientation are groove structures in micrometer scale. In this study, they were fabricated in titanium with a constant depth and varying width of 5 to 30 μm. Since titanium is the material of choice for orthopedic applications, the structures were tested with human fibroblasts and MG-63 osteoblasts [12].

A microscopic analysis revealed that the groove structures enabled cell orientation. This finding has already been observed in other studies [34, 120, 121]. In this work it was shown that the orientation was dependent on groove width and cell type. Fibroblasts reduced a parallel orientation on groove width larger than 15 μm, whereas osteoblasts were disarranged on width larger than 25 μm (Figure 29). Moreover, orientation was quantified with the standard derivation of parallel orientation, which gave more insights into the groove structure effects (Table 6). The correlation between orientation and cell type, was predicted beforehand [122]. Moreover, groove width was described to be the more critical parameter for cell orientation than groove depth [34]. It can be suggested that the selective control of cell orientation refers to different cell sizes. Fibroblasts are smaller in cell width and have an average L_c [μm] of 124.19 ± 6.36 and W_c [μm] of 20.49 ± 0.96, osteoblasts a L_c [μm] of 111.12 ± 4.12 and a W_c [μm] of 26.71 ± 1.3 (7.7). However, the cell shape is very dynamic and can be changed under certain conditions [93, 94]. It is dependent on the organization of the cytoskeleton which is controlled by adhesion to material surface. A more pronounced role of cell orientation appears to be the sensory guidance of cellular extensions such as filopodia, lamellipodia, and hemidesmosomen. They enable the binding to the the surface and the formation of focal adhesion complexes [93]. It can be supposed that the sensory guidance of cellular adhesion followed by cell orientation and organization of the cytoskeleton is cell specific. Further analysis are needed to clarify this sensory specificity of adhesion mechanism.

6.4.2 Selective control of cellular behavior by micro-, hierarchical nano- and micro- superimposed-, and nanostructures

The analysis of wettability of all structures revealed an increase of the water contact angel in comparison to the unstructured surface for all materials (Table 4). This measurement was carried out by Dipl.-Phys. E. Fadeeva at the Laser Zentrum Hannover e. V. (Germany) [41 - 43]. According to Cassie [46] this result refers to an incomplete wetting of the surface. Laser-induced changes in material chemistry could be excluded by energy dispersive X-ray spectroscopy (EDX) analysis which was performed by Dipl.-Phys. E. Fadeeva at the Laser Zentrum Hannover e. V. (Germany). Therefore, it was suggested that the wettability altering is just a property of topography [42]. In other words, the topography-dependent increase of the water contact angel reflected the reduction of surface area for contact [123]. This finding may be useful to predict cellular contact guidance [123].

Silicon, silicone elastomer and platinum are the material of choice for cochlea implants [31]. For this reason, the effectiveness for selective cell control by topographies produced in these materials were tested with human fibroblasts and SH-SY5Y neuroblastoma cells. Titanium is used for orthopedic applications [8]. Therefore, material effects on cellular behavior were analyzed with human fibroblasts and MG-63 osteoblasts.

On nano-structured platinum samples it was found, that fibroblasts decreased their cell growth when compared with unstructured platinum and the control. Furthermore, the reduction of proliferation correlated with the water contact angle of the samples: the bigger the decrease of surface area for contact, the lower the proliferation (Figure 33). A morphological analysis on spike structures produced in silicon and silicone reflected an increased roundness of fibroblasts and neuroblastoma cells given as the ratio of L_c/W_c (Figure 31 a). Furthermore, both cell types reduced the number of extensions (Figure 31 b). Hence, the morphology of the nuclei was not changed, in this case roundness of the cells did not correlate with either an increase of mitotic phases or an increase of cell death detectable through chromatin condensation. The same result was also observed for fibroblasts and osteoblasts cultivated on hierarchical nano- and micro- superimposed titanium structures. But in this case, material effects on morphology were cell type dependent. Fibroblasts showed a more rounded cell shape given as L_c/W_c and reduced the number of extensions, whereas osteoblasts were not affected (Figure 30). Furthermore, the structures enabled a selective control of cell proliferation. On the spike structures and on hierarchical nano- and micro- superimposed structures fibroblasts grew significantly slower than on the controls (Table 7, Figure 32 a). Simultaneously, neuroblastoma cells and osteoblasts proliferated at the same rate like under control conditions (Table 7, Figure 32 b). Comparable findings have already been described in literature [34, 37, 38, 124].

Since adhesion to material surface is considered to be the first step in biomaterial cell interaction [59], it can be suggested that surface topographies affect adhesion of the cells. All topographies presented a reduced surface area for contact. This means, the surface areas are reduced for adhesion ligands to associate on and are reduced for the whole cell body to attach. Possible effects on adhesion ligands refer to influences on the ligand concentration and conformation by the surface features. To clarifiy this idea, ones need to know, which adhesion ligands are used by the cells and if the cells react to varying ligand concentrations. After that the correlation between these ligands and topographies have to be addressed. Possible topographical effects on the adhesion of the whole cell body may be estimated by the knowledge of adhesion pattern, the localization, concentration, and type of the used adhesion receptor integrins. Furthermore, adhesion influences on the organization of the cytoskeleton in dependence of material properity needs to be analyzed. It is known that changes in cell morphology affect mechanical forces, which thereby could be contributed to changes in integrin-signaling [93, 118]. Since a selective cell control by the provided surface toporaphies was found, it can be suggested that the adhesion mechanism of the cells are cell specific. These facts point out that the knowledge of cell specific adhesion is the key step for finding biomaterials with perfect tissue integration.

6.5 Cell specificity of adhesion mechanism

The thesis that the selective cell control of materials in dependence of material chemistry and topography is caused by cell specific differences in adhesion mechanism, is up to now difficult to answer, since cell specificity of adhesion mechanism is poorly understood. Moreover, a detailed analysis of specificity is complicated, as adhesion mechanism are very complex and many different factors are involved. For instance, Hehlgans [56] described at least 24 different integrin receptors, Tzu [60] and Heino [74] mentioned a huge variety of adhesion ligands like 16 different laminins and 29 different collagens.

In the past, the investigations of adhesion mechanism were concentrated on biochemical and molecular biological measurements. With the help of antibodies, receptors can be detected and localized [125]. The use of ligand-functionalized biomaterials is a promising method to analyze ligand-receptor interactions [126, 127]. However, common techniques have several disadvantages. First, they are costly. Second, the huge range of possible receptors and ligands and the missing knowledge about cell specificity makes it hard to choose the right antibodies. Third, no characterization of biophysical functions such as kinetic observations is possible. For this reason, two new methods were introduced in this work which can be used to indicate disparities of adhesion mechanism without knowing the detailed factors such as adhesion ligands and adhesion receptors that may be involved. These methods are analysis of adhesion kinetic and adhesion pattern.

To figure out cell specific differences of adhesion mechanism, the effects of adhesion ligands such as laminin, fibronectin, collagen type I, and vitronectin were analyzed. Besides taking into account ligand concentration dependencies, it was investigated, whether the cells respond to all ligands or not, whether some ligands play a more important role than others or a comparable one. Cellular responses to the ligands were characterized via adhesion time, adhesion pattern, morphology, proliferation, and gap junction coupling.

Two different experimental setups were choosen for the investigations. First, a shortterm analysis was performed with cell culture media that did not include serum. Since serum consists of ligands like fibronectin and vitronectin [78, 128] these factors may be in contradiction with the specificity of the experiments using ligand-coated substrates. Second, a longterm analysis was carried out with serum-including cell culture media to provide growth factors and hormons needed for cell growth and to inhibit apoptotic reactions [129, 130]. Furthermore, differences in adhesion mechanism are thought to be more pronounced after a longer cultivation time [131].

All measurements were performed with various cell types such as human fibroblasts, GM-7373 endothelial cells, SH-SY5Y neuroblastoma cells, HaCaT keratinocytes, MG-63 osteoblasts, and A10 smooth muscle cells to provide a wide range of tissues that are of interest for biomedical applications.

6.5.1 Adhesion time and pattern are cell specific

In this study a novel method was introduced that allows a documentation of biophysical functions of adhesion mechanism. By counting the cell densities of nonadherent cells in one hour time intervals, one is able to estimate the novel parameter adhesion time A_T, defined as the time needed until half of nonadherent cells at $t = 0$ h attached to the surface. To exclude possible side effects like detached cells from the surface, the measurement is restricted to a total measuring time of five hours. It can be applied to any adherend cell type and any material of interest in a costeffective and fast manner. Therefore, this method provides an insight into understanding adhesion mechanism in general, but it could also find its application in the field of biomedicine and tissue engineering to investigate the influence of biomaterials on cellular behavior.

It was found, that the cells adhere to the control surface with a defined speed. The fastest adhesion time was observed for osteoblasts followed by keratinocytes and fibroblasts, neuroblastoma cells, endothelial cells, and smooth muscle cells (Table 8). The parameter adhesion time is cell specific.

However, from the physical point of view the calculation of adhesion time A_T does not include all parameters needed for cellular adhesion. The physics of cell adhesion

Discussion

contains many different steps and a variety of intermolecular forces. Theoretically, it depends on the binding probability, the binding strength, and the interaction area. To elaborate a theory, it is hard to describe living cells as a physical object related to the diversity and dynamics of cell components, plasma membrane, cell surface charge, membrane viscosity, and diffusion of molecules. Several efforts were undertaken for a physical approach to calculate adhesion [132]. However, a detailed and not generalized formula is still missing which includes all possible paramters such as Van der Waal forces, electrostatic and hydrophobic interactions, inter/- and intracellular interactions, surface mechanic and chemistry, cell mechanic, adsorption kinetic of adhesion ligands, change in the conformation, density, localization, expression, and affinity of adhesion receptors and ligands, cell specificity, and others [55, 133]. Therefore, the calculation of adhesion time A_T may be a good alternative for the physical description of cellular adhesion mechanism.

The surface reflectance interference contrast (SRIC) technique can be used to visualize adhesion pattern of cells. By reflecting light at the interface between cells and cultivation surface, close adhesion contacts appear dark, big cell-surface distance bright. According to Adams [93] the investigated pattern refer to cell-matrix contacts formed by the binding of adhesion ligands to integrins.

Under control conditions neuroblastoma cells, keratinocytes, and osteoblasts formed many small focal contacts to the surface. On the contrary, fibroblasts, endothelial cells, and smooth muscle cells seemed to adhere with the whole cell body (Figure 34). These observation demonstrate that the adhesion pattern of the investigated cell types are different.

Both methods such as analysis of adhesion kinetic and adhesion pattern pointed out cell specificity. Therefore, it can be concluded that the adhesion mechanism of analyzed cell types have to be different. Theoretically, these disparities can be caused by (a) the use of different adhesion ligands from the extracellular matrix and / or different adhesion receptors, (b) a comparable range of ligands and receptors, but variations in priorities, densities, concentrations, and localizations, and (c) differences in their dynamic availability and interactions, also with unknown factors.

6.5.2 Cell specific ligand priority ranking in dependence of the ligand concentration

In presence of the ligands, the cell specific adhesion time was accelerated ligand independently (Figure 43). With the help of the standard procedure, all cells reduced their adhesion time A_T to less than one hour in the presence of the ligands, whereas on the control the values were ranged between two and five hours (Table 8). The acceleration can be explained by the fact, that surfaces were already 'attractive' for cell adhesion, since they were coated with ligands. Under control conditions the adhesion ligands coming from the serum-containing cell culture media have to adsorp on material surface. Furthermore, the ligands have to form the right conformation which enables the binding to the cell. Both effects take time and are excluded by using ligand-coated substrates (Figure 43). On ligand-coated surfaces the cells were able to attach to the ligands directly. This association is the key step for cell binding to the surface and adhesion [59].

Since an acceleration of adhesion time was found cell type independently, it can be concluded that the cells respond to all ligands. Theoretically, an acceleration of adhesion time can also be transferred to specific acceleration of the doupling time, an increase of cell elongation, an increase in the number of extensions, and influences on gap junction coupling [59]. With respect to all performed measurements, a cell specific ligand priority ranking with the maximum ligand concentration was found in comparison to the control.

Human fibroblasts adhered the fastest (Table 10), were most elongated (Figure 37 a), formed more extensions (Figure 37 b), reduced the doubling time (Table 12), and increased gap junction coupling (Table 13) on fibronectin. The best responses to fibronectin were followed by the responses to collagen, laminin, and at last vitronectin. Fibroblasts have been shown to interact with fibronectin [134].

GM-7373 endothelial cells were found to prefer vitronectin. This finding was supported by the fastest adhesion (Table 10), cell shape (Figure 38 a), the highest amount of extensions (Figure 38, Figure 37b), the reduced the doubling time (Table 12), and increased gap junction coupling (Table 13). After vitronectin the ranking was laminin, collagen, and fibronectin. For endothelial cells positive interactions with vitronectin and negative with fibronectin have already been described [78, 79, 135].

Discussion

SH-SY5Y neuroblastoma cells responded best to laminin as seen for adhesion time (Table 10), cell elongation (Figure 39 a), the number of extensions (Figure 39, Figure 37b), and the reduced the doubling time (Table 12). Then the ligand order was vitronectin, followed by fibronectin and collagen. Neuroblastoma cells were shown to prefer laminin, the major molecule of the basallamina [62].

The responses of HaCaT keratinocytes were comparable with fibroblasts. They also preferred fibronectin followed by collagen. But the most reduced reponses were found for laminin and not vitronectin. This finding is attested by the smallest adhesion time (Table 10), the ratio of L_c/W_c (Figure 40 a), the highest amount of extensions (Figure 40, Figure 37b), the reduced the doubling time (Table 12), and increased gap junction coupling (Table 13). It was demonstrated that keratinocytes bind to collagen type I and RGD sequence-including ligands [61, 136].

MG-63 osteoblasts preferred vitronectin and collagen as seen for the adhesion time (Table 10), cell elongation (Figure 48 a), the highest amount of extensions (Figure 48, Figure 37b), and the reduced the doubling time (Table 12). This ranking was completed by laminin followed by fibronectin.

The ligand priority ranking of A10 smooth muscle cells was in the order laminin, fibronectin, collagen, and vitronectin examined by the adhesion time (Table 10), cell dilation (Figure 49 a), the highest amount of extensions (Figure 49, Figure 37b), the reduced the doubling time (Table 12), and increased gap junction coupling (Table 13). It was reported that smooth muscle cells refer to laminin and collagen [75, 137, 138].

Cell type independently, a reduced concentration of the preferred ligand correlated with a decrease of adhesion kinetic. Simultaneouly, a reduced concentration of the not preferred ligand with an increase of adhesion kinetic. Only neuroblastoma cells did not react to varying ligand concentrations as seen for adhesion time and proliferation (Table 10, Table 12). In literature a concentration dependency has only been described for the presence of fibronectin [139].

Ligands effects are also transferred on cell proliferation. It is formerly known that cell cycle progression is controlled by integrin-mediated adhesion to the surface [56, 86, 89, 91]. An increase in the doubling time may correlate with the saved time caused by an increase of adhesion time (Table 10). However, integrin signaling also stimulates regulatory molecules of the cell cycle like cyclins and cyclin-dependent kinase (CDK)

[86, 91]. In what manner the used ligands stimulate cell cycle progression needs further analysis with respect to cyclin and CDK functions.

The preference of the ligand referred to an increase in the elongation of the cells given as the ratio of L_c/W_c and an increase in the number of extensions [93]. Poor adhesion correlates with a rounded shape and a reduced number of extension. Theoretically, a rounded shape could also be caused by mitotic phases and chromatin condensation. Both phenomena were excluded by analyzing the nucleus dilation given as the ration of L_n/W_n, so that all changes in cell morphology on the ligands refer to adhesion.

The analysis of gap junction coupling did not point out a clear correlation between the ligand priority ranking and varying ligand concentrations (Table 13). A detailed explanation for the observed changes in gap junction coupling are still missing and need further analysis. Several studies suggested that the extracellular matrix composition affects gap junction coupling [104, 140]. These components may modulate connexin expression and posttranslational modifications. Imbeault [101] found that laminin upregulates, downregulates, and changes the localization of specific connexins in neural progenitor cells. Furthermore, the hemichannel activity may be suppressed on laminin. Other studies suggested that extracellular matrix effects on gap junction coupling were rather caused by matrix-controlled mechanical forces on the cytoskeleton, which thereby regulate the open kinetic and probability of gap junction channels [141, 142]. Another probability may be, that changes in gap junction coupling do not refer to connexin expressions and channel activity, but to the second messengers that are distributed over gap junctions to neighboring cells. For instance, endothelial cells probably require a higher cytosolic concentration of Ca^{2+} by the presence of vitronectin [78]. The binding of fibroblasts to RGD-sequences was also thought to be Ca^{2+}-dependent [143]. However, cell and ligand specific influences on connexins, gap junction activity, and second messengers such as Ca^{2+} are up to now poorly understood.

In this work, the analysis of cell responses to adhesion ligands excluded ligand influences on integrins and other components involved in integrin signaling. However, these measurements are necessary for a more detailed understanding of cell specific differences in adhesion mechanism.

The investigations of adhesion pattern give an insight in ligand effects on integrins, since these pattern characterize cell surface distances. Close distances appear dark with respect to the surface reflectance interference contrast technique, and indicate the formation of focal contacts composed of integrins [93]. It was found that on the control human fibroblasts, GM-7373 endothelial cells, and A10 smooth muscle cells formed close contacts with the entire cell body (Figure 34 a, b, f). On the contrary, SH-SY5Y neuroblastoma cells, HaCaT keratinocytes, and MG-63 osteoblasts formed many small focal contacts (Figure 34 c, d, e). Therefore, it can be supposed that the localization and densities of the integrins and the focal contacts occur in a cell specific manner. To quantify the cell surface distances a new method was introduced by creating a histogram that represents the distribution and average of the gray values within the cell body.

Cultivated on the adhesion ligands, the adhesion pattern were changed as the contacts were increased or decreased. Fibroblasts formed closer contacts on fibronectin and laminin after the shortterm method, later also on collagen (Table 9). On vitronectin it was always increased (Table 11). Endothelial cells adhered closer on vitronectin and laminin, and less on fibronectin and collagen in both setups (Table 9, Table 11). Closest cell surface distances occurred for neuroblastoma cells on laminin. The average gray value was increased on the other ligands (Table 9, Table 11). For keratinocytes the values were in the same range, but on vitronectin it was reduced after the shortterm and increased after the longterm analysis (Table 9, Table 11). Osteoblasts were not significantly affected by the ligands after the shortterm method (Table 9). After the longterm method the distances were increased (Table 11). The cell surface distances of smooth muscle cells were all comparable and did not show any differences (Table 9, Table 11).

All histograms presented maxima standing for gray values that occurred more often (Figure 35). As the values correlate with the distance between the cells and the surface, it can be suggested that the maxima represent defined binding distances of the used adhesion ligand and receptor. As the histograms included several maxima, the cells may form a wide range of binding motives to the surface. This thesis is supported by the fact that there is an overlap in specificity and affinity, with many integrins capable of binding to more than one protein, whereas proteins can act as ligands for more than one integrin. At least nine integrins have been described to bind laminin such as $\alpha_1\beta_1$, $\alpha_2\beta_1$, $\alpha_3\beta_1$, $\alpha_5\beta_1$, $\alpha_6\beta_1$, $\alpha_6\beta_4$, $\alpha_v\beta_3$, $\alpha_v\beta_5$, and $\alpha_7\beta_1$ [57, 60 - 65]. Fibronectin can bind to $\alpha_3\beta_1$, $\alpha_4\beta_1$, $\alpha_5\beta_1$, $\alpha_8\beta_1$, $\alpha_v\beta_1$, $\alpha_v\beta_3$, $\alpha_v\beta_6$, and $\alpha_4\beta_7$ integrins [62, 66, 69 - 73]. The binding motif of collagen is recognized by $\alpha_1\beta_1$, $\alpha_2\beta_1$, $\alpha_{10}\beta_1$, $\alpha_{11}\beta_1$, and $\alpha_v\beta_8$ integrins [61, 62, 69,

71, 75, 76]. Integrins such as $\alpha_5\beta_1$, $\alpha_V\beta_1$, $\alpha_V\beta_3$, and $\alpha_V\beta_5$ bind to vitronectin [69, 71, 79]. Furthermore, there is a variety of integrins that are expressed in the cell types. For instance, for fibroblasts $\alpha_1\beta_1$, $\alpha_2\beta_1$, $\alpha_3\beta_1$, $\alpha_5\beta_1$ integrins were identified [143, 144], for neuroblastoma cells $\alpha_1\beta_1$, $\alpha_3\beta_1$, $\alpha_5\beta_3$, $\alpha_V\beta_3$ integrins [53, 62], for endothelial cells $\alpha_2\beta_1$, $\alpha_5\beta_3$, $\alpha_6\beta_4$, $\alpha_V\beta_3$ integrins [69, 145, 146], for keratinocytes $\alpha_2\beta_1$, $\alpha_3\beta_1$, $\alpha_5\beta_1$, and others [61, 147], for osteoblasts the subunits α_1, α_2, α_3, α_4, α_5, α_6, α_V, β_1, β_3, β_5 [72] and for smooth muscle cells $\alpha_1\beta_1$, $\alpha_2\beta_1$, $\alpha_5\beta_1$, $\alpha_V\beta_3$ [71, 148].

However, further analysis are needed to clarify this topic. For instance, it needs to be examined which integrins are expressed in the used cell types, where these integrins are localized, whether expression and localization change in dependence of the cultivation time and conditions. Furthermore, which integrins respond to the ligands, especially what binding distances are achieved. Nevertheless, the quantification of the adhesion pattern revealed cell and ligand specifc differences. But imaging and quantification of adhesion pattern need to be improved as well. The first disadvantage relies on the fact, that this method is only applicable for transparent and thin surfaces. By this means, it can not be used for biomaterials and surface topographies. The second disadvantage refers to the quantified distances which can only be measured in pixel scale. A huge benefit would be the possible calculation in micro- or nanometer scale.

Small differences occurred with respect to both experimental setups. The quantification of cell morphology revealed, that the cultivation under serum-free conditions limited to five hours, caused a more rounded cell shape and basically reduced the number of extensions (Figure 37 to Figure 42, Figure 44 to Figure 49). A closer look to the adhesion pattern revealed, that after the shortterm experiment all cell types formed bigger distances to the surface (Table 9, Table 11). Whether both effects correlated with the reduced cultivation time or the missing serum, needs further analysis.

6.6 The selective cell control of biomaterials was caused by cell specific differences in adhesion mechanism

6.6.1 Correlation between hydrophobicity and adhesion

In dependence of the hydrophobic character of the material, the used cell types could be separated into two classes: cells preferring hydrophilic materials like human fibroblasts versus cells preferring hydrophobic materials like GM-7373 endothelial cells and SH-SY5Y neuroblastoma cells (6.3.1). Comparable effects have been described elsewhere, but the explanation was missing [111 - 113].

Probably, the essential step of the selective cell control by material hydrophobicity refers to ligand adsorption to material surfaces. Several studies demonstrated that material chemistry guides protein adsorption correlating with the protein orientation, mobility, density, and conformation [137, 149, 150]. Whereas the kinetic of ligand adsorption does not depend on material hydrophobicity, the achieved conformation does [151]. Most of these studies have focussed on the ligand fibronectin. It was found that fibronectin presents a conformation on hydrophobic materials that hides the important integrin recognition motif RGD-sequence [66, 152]. By this means, the integrin receptors cannot bind to the ligands, and thereby inhibit the adhesion to the surface. Furthermore, it was shown that the secondary structure of fibronectin is denaturated [135, 153]. On the contrary, the ligand vitronectin does not change its conformation [154]. Even though little is known about hydrophobicity effects on laminin and collagen, with these findings it can be concluded that cells using primary fibronectin-integrin binding adhere on hydrophilic materials and do not adhere on hydrophobic materials. This conclusion is supported by the found cell specific ligand priority ranking. Fibroblasts which primarily use fibronectin preferred hydrophilic materials. In contrast to fibroblasts, endothelial cells and neuroblastoma cells primarily used vitronectin and laminin. It can be supposed that these ligands rather show positive interactions with hydrophobic materials, since these cells did not respond well to hyrophilic materials. Whether material hydrophobicity also influenced ligand concentrations, needs further analysis.

However, the found cell specific ligand priority ranking may be helpful to predict cell responses in dependence of material hydrophobicity. It can be predicted that cells using primary fibronectin attachment will adhere well on hydrophilic materials and not on hydrophobic materials. The cell types having a different preferred ligand may show the opposite effect. This ranking can also be applied for a biological functionalization to combine biomaterials with adhesion ligands or binding sequences, since now the preferred and cell specific ligands are known. Therefore, this ligand priority ranking facilitates material research and functionalization for future biomedical applications.

6.6.2 Correlation between topography and adhesion

On surface topographies presenting a reduced surface area for contact, human fibroblasts were always inhibited, and SH-SY5Y neuroblastoma cells and MG-63 osteoblasts not (6.4). Comparable findings have already been described in literature, but a clear explanation was missing [34, 37, 38, 124].

All surface topographies that were used showed a reduced surface area for contact (Table 4). In other words the samples showed a reduced surface area for ligand adsorption. It was described that fibronectin adsorption correlates with the surface area. On smaller surface areas, a reduced concentration was achieved [155]. Furthermore, the ligand vitronectin is with 15 nm length a lot smaller than fibronectin, and therefore the concentration changes may be not so pronounced on the topographies [156]. With respect to the found cell sensitivity in dependence of ligand concentrations, this may be the first approach for explanation. This thesis is supported by the fact, that neuroblastoma cells were the only cell type that did not react to varying ligand concentrations. For this reason they were probably not negatively affected by the surface topographies. Even though osteoblasts reacted to varying ligand concentrations, they may have been not negatively affected because they primary use vitronectin. Concerning fibronectin it was also found that the achieved concentration on the surface directly influences protein conformation. With small concentrations the protein cannot unfold and thereby hides the RGD-cell binding domain [151]. If this is the case, fibroblasts using primary fibronectin cannot bind to surface topographies with a reduced surface area that directly decreases fibronectin adsorption. As neuroblastoma cells and osteoblasts prefer laminin and vitronectin, this may be a second possible explanation. This is supported by the fact that vitronectin does not react to conformational changes [154]. However, further analysis have to focus on topographical effects on ligand adsorption, ligand concentration, and conformation.

The produced surface topographies did not only present a reduced surface area for contact for protein adsorption, but also a reduced surface contact area for the total cell body. This may be also a critical paramter for cell adhesion. The analysis of adhesion pattern revealed, that fibroblasts adhere with the total cell body under control conditions, whereas neuroblastoma cells and osteoblasts only form focal contacts (Figure 34). Therefore, it can be concluded that the surface structures do not offer enough contact area for fibroblasts to adhere and proliferate, whereas neuroblastoma cells and osteoblasts are not negatively affected [42]. Since the adhesion pattern refer to ligand-receptor distances, it has to be analyzed in the future, if the expression and localization of integrins are changed on the structures. Furthermore, the surface structures influenced cell morphology, especially the shape of fibroblasts (Figure 30 b, Figure 31 a). Changes in cell morphology affect mechanical forces, which thereby could be contributed to changes in integrin-signaling [96, 118]. In what manner the investigated cell types were selectively and mechanically affected by surface structures needs further analysis.

To predict cell responses to topographical features, two analysis were shown to be helpful. First, wettability changes caused by structuring indicate a decrease or increase of surface area for contact. Second, adhesion pattern demonstrate if the cells adhere with the whole cell body or form small focal contacts. Cells needing large contacts to the surface are probably negatively affected by structures presenting a reduced surface area for contact. The combination of both analysis facilitates the material search and functionalization for future biomedical applications.

6.7 Conclusions

This work addressed a wide platform for biomedical interest. Three different laser technologies were introduced that enable the precise and controlled design of three-dimensional structures (by Dr. A. Ovsianikov), cell transport (by Dipl.-Ing. M. Güne and Dr. L. Koch), and surface topographies in micro- and nanometer scale (by Dipl.-Phys. E. Fadeeva). The negative replication technique also revealed several advantages (by Dipl.-Phys. E. Fadeeva). Material aging was a possible tool to overcome the toxicity of the applied photoinitiators. All methods were carried out in the Laser Zentrum Hannover e. V. (Germany).

A comparative cell study was performed to figure out cell responses in dependence of the material with variable properties, scaffolds, cell transport, and surface topographies. The cell transport did not negatively affect cellular behavior. Cell localization on three-dimensional scaffolds was dependent on scaffold size dimensions. A selective cell control in dependence of material hydrophobicity, crosslinking density, and surface topographies with a reduced surface area for contact was found. It was suggested that this control is caused by cell specific differences in adhesion mechanism. For this purpose, novel methods such as analysis of adhesion time and pattern were established that provide new insights in this direction. Furthermore, specificity of adhesion mechanism with respect to influences of different adhesion ligands also in dependence of their concentration was analyzed. With this method a cell specific ligand priority ranking was found. Moreover, the cells reacted to varying ligand concentrations. These findings were used to explain the observed selective cell control. Furthermore, these findings are promising to predict cell responses to various materials and facilitate material search and functionalization for future biomedical applications.

6.8 Future perspective

In the future other materials have to be analyzed with respect to cell responses and their possibility of laser manufacturing. Disadvantages of some materials related to mechanical stability have to be overcome in order to fabricate three-dimensional structures. The laser-induced forward transfer has to be applied to produce multiple cell layers and to coat directly three-dimensional scaffolds with cells. By this means, established microscopic techniques have to be improved that enable cell documentation in three dimensions. Furthermore, cell behavior in three dimensions has to be investigated. Surface structuring has to be performed on implants directly, as well as on other materials also with different topographical features. The cell experiments have to be widened on other cell types, and especially primary cell lines and stem cells. The analysis of adhesion pattern needs improvements due to the accurate correlation between the gray scales and cell-surface distances in micro- or nanometer scale. Concerning the found cell specificity of adhesion mechanism, further questions still remained open such as the role of integrins, sensory guidance, cytoskeletal elements like vinculin and others, mechanical forces, members of integrin signaling, ligand effects on cell cycle progression, and gap junction coupling. All investigations have also to be carried out in dependence of biomaterials with various properties such as hydrophobicity, stiffness, and topography. It needs to be analyzed in what cell specific manner the materials affect adhesion ligands and thereby the total adhesion mechanism.

7 Attachment
7.1 References

[1] **Geetha, M. et al.** „Ti based biomaterials, the ultimate choice for orthopaedic implants – a review", *Progress in Material Science*, 54: 397-425, 2009

[2] **Pye, A. D. et al.** „A review of dental implants and infection", *Journal of Hospital Infection*, 72: 104-110, 2009

[3] **Temenoff, J. S. et al.** „Review: tissue engineering for regeneration of articular cartilage", *Biomaterials*, 21: 431-440, 2000

[4] **Shinohara, T. et al.** „Neurotrophic factor intervention restores auditory function in deafened animals", *PNAS*, Vol. 99, No. 3: 1657-1660, 2002

[5] **Yang, F. et al.** „The effect of incorporating RGD adhesive peptide in polyethylene glycol diacrylate hydrogel on osteogenesis of bone marrow stromal cells", *Biomaterials*, 26: 5991-5998, 2005

[6] **Bacakova, L. et al.** „Molecular mechanism of improved adhesion and growth of an endothelial cell line cultured on polystyrene implanted with fluorine ions", *Biomaterials*, 21: 1173-1179, 2000

[7] **Lee, K. Y. et al.** „Hydrogels for tissue engineering", *Chemical Reviews*, Vol. 101, No. 7: 1869-1879, 2001

[8] **Buford, A. et al.** „Review of wear mechanism in hip implants: paper I - general", *Materials and Design*, 25: 385-393, 2004

[9] **Slonaker, M. et al.** „Review of wear mechanism in hip implants: paper II - ceramics", *Materials and Design*, 25: 395-405, 2004

[10] **Kurella, A. et al.** „Review paper: Surface modification for bioimplants: the role of laser surface engineering", *Journal of Biomaterials Applications*, 20: 5-50, 2005

[11] **Brandl, F. et al.** „Rational design of hydrogels for tissue engineering: impact of physical factors on cell behavior", *Biomaterials*, 28: 134-146, 2007

[12] **Cooper, L. F.** „Biologic determinants for bone formation for osseointegration: clues for future clinical improvements", *Journal of Prosthetic Dentistry*, 80: 439-449, 1998

[13] **Papavasiliou, G. et al.** „Three-dimensional patterning of poly(ethylene glycol) hydrogels through surface-initiated photopolymerization", *Tissue Engineering*, Vol. 14, No. 2: 129-140, 2008

[14] **Alsberg, E. et al.** „Engineering growing tissues", *Applied Biological Science*, Vol. 99, No. 19: 12025-12030, 2002

[15] **Drury, J. L. et al.** „Hydrogels for tissue engineering: scaffold design variables and applications", *Biomaterials*, 24: 4337-4351, 2003

[16] **Jones, J. R.** „Observing cell response to biomaterials", *Materials Today*, Vol. 9, No. 12: 34-43, 2006

[17] **Fedorovich, N. E. et al.** „Hydrogels as extracellular matrices for skeletal tissue engineering: state-of-the-art and novel application in organ printing", *Tissue Engineering*, Vol. 13, No. 8: 1905-1925, 2007

[18] **Moscato, S. et al.** „Morphological features of ovine embryonic fibroblasts cultured on different scaffolds", *Journal of Biomedical Material Research*, 2005

[19] **Cukierman, E. et al.** „Taking cell-matrix adhesions to the third dimension", *Science*, 294: 1708-1712, 2001

[20] **Cukierman, E. et al.** „Cell interactions with three-dimensional matrices", *Current Opinion in Cell Biology*, 14: 633-639, 2002

[21] **Schlie, S. et al.** „ Three-dimensional cell growth on structures fabricated from Ormocer® by two-photon polymerisation technique", *Journal of Biomaterials Applications*, Vol. 22, No. 3: 275-287, 2007

[22] **Ovsianikov, A et al.** „Two-photon polymerization technique for microfabrication of CAD-designed 3D scaffolds from commercially available photosensitive materials", *Journal of Tissue Engineering and Regenerative Medicine*, 1: 443-447, 2007

[23] **Jeon, O. et al.** „Photocrosslinked alginate hydrogels with tunable biodegradation rates and mechanical properties", *Biomaterials*, 30: 2724-2734, 2009

[24] **Dijk-Wolthuis, W. N. E. et al.** „Degradation kinetics of methacrylated dextrans in aqueous solution", *Journal of Pharmaceutical Science*, Vol. 86, No. 4: 413-417, 1997

[25] **Franssen, O. et al.** „Degradable hydrogels: controlled release of a model protein from cylinders and microspheres", *Journal of Controlled Release*, 60: 211-221, 1999

[26] **Barron, J. A. et al.** „Biological laser printing: a novel technique for creating heterogenous 3-dimensional cell patterns", *Biomedical Microdevices*, 6: 139-147, 2004

[27] **Lewis, J. et al.** „Direct writing in three dimensions", *Materialstudy*, 32-39, 2004

[28] **Kaji, T. et al.** „Nondestructive micropatterning of living animal cells using focused femtosecond laser-induced impulsive force", *Applied Physics Letters*, 91, DOI: 10.1063/1.2753103, 2007

[29] **Loesberg, W. A. et al.** „The threshold at which substrate nanogroove dimensions may influence fibroblast alignment and adhesion", *Biomaterials*, 28: 3944-3951, 2007

[30] **Anderson, J. M.** „Biological responses to materials", *Annu. Rev. Mater. Res.*, 31: 81-110, 2001

[31] **Reich, U. et al.** „Differential fine-tuning of cochlear implant material-cell interactions by femtosecond laser microstructuring", *Mater Res Part B: Appl Biomater*, 87B: 146-153, 2008

[32] **Barbucci, R. et al.** „Micropatterned surfaces for the control of endothelial cell behaviour", *Biomolecular Engineering*, 19: 161-170, 2002

[33] **Dalby, M. J. et al.** „Rapid fibroblast adhesion to 27 nm high polymer demixed nano-topography", *Biomaterials*, 25: 77-83, 2004

[34] **Flemming, R. G . et al.** „Effects of synthetic micro- and nano-structured surfaces on cell behavior", *Biomaterials*, 20: 573-588, 1999

[35] **Weiss, P. et al.** „Shape and movement of mesenchyme cells as functions of the physical structure of the medium", *Proceedings of the National Academy of Science of the United States of America*, 38: 264-280, 1952

[36] **Curtis, A. et al.** „Nanotechniques and approaches in biotechnology", *Trends in Biotechnology*, Vol. 19, No. 3: 97-101, 2001

[37] **Yim, E. K. F. et al.** „Nanopattern-induced changes in morphology and motility of smooth muscle cells", *Biomaterials*, 26: 5405-5413, 2005

[38] **Choi, C. H. et al.** „Cell interaction with three-dimensional sharp-tip nanotopography", *Biomaterials*, 28: 1672-1679, 2007

[39] **Lim, J. Y. et al.** „Cell sensing and response to micro- and nanostructured surfaces produced by chemical and topographical patterning", *Tissue Engineering*, 13: 1879-1891, 2007

[40] **Coen, M. C. et al.** „Modification of the micro- and nanotopography of several polymers by plasma treatment", *Applied Surface Science*, 207: 276-286, 2003

[41] **Fadeeva, E. et al.** „Femtosecond laser-induced surface structures on platinum and their effects on surface wettability and fibroblast cell proliferation", *Contact Angle, Wettability and Adhesion*, Vol. 6: 163-171, 2009

[42] **Schlie, S. et al.** „Femtosecond laser fabricated spike structures for selective control of cellular behavior", *Journal of Biomaterials Applications*, JBA-09-0074.R1, 2009

[43] **Fadeeva, E. et al.** „The hydrophobic properties of femtosecond laser fabricated spike structures and their effects on cell proliferation", *Physica Status Solidi*, A: 1-4, 2009

[44] **Kannan, R. Y. et al.** „Principles and applications of surfaces analytical techniques at the vascular interface", *Journal of Biomaterials Applications*, 21: 5-31, 2006

[45] **Wenzel, R. N.** „Resistance of solid surfaces to wetting by water", *Ind. Eng. Chem.*, 28: 988-994, 1936

[46] **Cassie, A. B. D. et al.** „Wettability of porous surfaces", *Trans. Faraday Soc.*, 40: 546-551, 1944

[47] **El-Amin, S. F. et al.** „Extracellular matrix production by human osteoblasts cultured on biodegradable polymers applicable for tissue engineering", *Biomaterials*, 24: 1213-1221, 2003

[48] **Powell, H. M. et al.** „Nanotopographic control of cytoskeletal organization", *Langmuir*, 22: 5087-5094, 2006

[49] **Lim, J. Y. et al.** „The regulation of integrin-mediated osteoblast focal adhesion and focal adhesion kinase expression by nanoscale topography", *Biomaterials*, 28: 1787-1797, 2007

[50] **Neff, J. A. et al.** „Surface modification for controlled studies of cell-ligand interactions", *Biomaterials*, 20: 2377-2393, 1999

[51] **Clark, E. A. et al.** „Ras activation is necessary for integrin-mediated activation of extracellular signal-regulated kinase 2 and cytosolic phospholipase A_2 but not for cytoskeletal organization", *Journal of Biological Chemistry*, Vol. 271, No. 5: 14814-14814, 1996

[52] **Aplin, A. E. et al.** „Cell adhesion molecules, signal transduction and cell growth", *Current Opinion in Cell Biology*, 11: 737-744, 1999

[53] **Bozzo, C. et al.** „Soluble integrin ligands and growth factors independently rescue neuroblastoma cells from apoptosis under nonadherent conditions", *Experimental Cell Research*, 237: 326-337, 1997

[54] **Bonfocco, E. et al.** „β_1 integrin antagonism on adherent, differentiated human neuroblastoma cells triggers an apoptotic signaling pathway", *Neuroscience*, Vol. 101, No. 4: 1145-1152, 2000

[55] **García, A. J.** „Get a grip: integrins in cell-biomaterial interactions", *Biomaterials*, 26: 7525-7529, 2005

[56] **Hehlgans, S. et al.** „Signaling via integrins: implications for cell survival and anticancer strategies", *Biochimica et Biophysica Acta*, 1775: 163-180, 2007

[57] **Adair-Kirk, T. L. et al.** „Fragments of extracellular matrix as mediators of inflammation", *The International Journal of Biochemistry & Cell Biology*, 40: 1101-1110, 2008

[58] **Gumbiner, B.** „Cell adhesion: the molecular basis of tissue architecture and morphogenesis", *Cell*, Vol. 84: 345-357, 1996

[59] **Miyamoto, S. et al.** „Integrin function: molecular hierarchies of cytoskeletal and signaling molecules", *Journal of Cell Biology*, 131: 791-805, 1995

[60] **Tzu, J. et al.** „Bridging structure with function: structural, regulatory, and developmental role of laminins", *International Journal of Biochemistry and Cell Biology*, 40: 199-214, 2008

[61] **Schafer, I. A. et al.** „Human keratinocytes cultured on collagen gels form an epidermis which synthesizes bullous pemphigoid antigens and $\alpha_2\beta_5$ integrins and secretes laminin, type IV collagen, and heparan sulfate proteoglycan at the basal cell surface", *Experimental Cell Research*, 195: 443-457, 1991

[62] **Letourneau, P. C. et al.** „ Extracellular matrix and neurite outgrowth", *Current Biology*, 2: 625-634, 1992

[63] **García, A. J. et al.** „Integrin-fibronectin interactions at the cell-material interface: initial integrin binding and signaling", *Biomaterials*, 20: 2427-2433, 1999

[64] **Bolcato-Bellemin, A.-L. et al.** „Laminin α_5 chain is required for intestinal smooth muscle development", *Developmental Biology*, 260: 376-390, 2003

[65] **Su, L. et al.** „Neural stem cell differentiation is mediated by integrin β_4 in vitro", *The International Journal of Biochemistry & Cell Biology*, 41: 916-924, 2009

[66] **Kowalczynska, H. M. et al.** „Fibronectin adsorption and arrangement on copolymer surfaces and their significance in cell adhesion", *Journal of Biomedical Material Research*, 72A: 228-236, 2005

[67] **Hynes, R. et al.** „Fibronectins: multifunctional modular glycoproteins", *The Journal of Cell Biology*, 95: 369-377, 1982

[68] **Craig, D. et al.** „Structural insights into how the MIDAS ion stabalizes integrin binding to an RGD peptide under force", *Structure*, 12: 2049-2058, 2004

[69] **Chen, C. S. et al.** „Geometric control of cell life and death", *Science*, 276: 1425-1428, 1997

[70] **Altankov, G. et al.** „Morphological evidence for different fibronectin receptor organization and function during fibroblast adhesion on hydrophilic and hydrophobic glass substrata", *J. Biomater. Sci. Polymer Edn*, Vol. 8, No. 4: 721-740, 1997

[71] **Blaschke, F. et al.** „Hypoxia activates β_1 – integrin via ERK 1/2 and p38 MAP kinase in human vascular smooth muscle cells", *Biochem and Biophys Research Comm*, 296: 890-896, 2002

[72] **Siebers, M. C. et al.** „Integrins as linker proteins between osteoblasts and bone replacing materials. A critical review", *Biomaterials*, 26: 137-146, 2005

[73] **Salsmann, A. et al.** „RGD, the Rho'd cell spreading, *European Journal of Cell Biology*, 85: 249-254, 2006

[74] **Heino, J. et al.** „Evolution of collagen-based adhesion systems", *The International Journal of Biochemistry & Cell Biology*, 41: 341-348, 2009

[75] **Koyama, H. et al.** „Fibrillar collagen inhibits arterial smooth muscle proliferation through regulation of cdk2 inhibitors", *Cell*, 87: 1067-1078, 1996

[76] **Kreis, S. et al.** „The intermediate filament protein vimentin binds specifically to a recombinant integrin α_2/β_1 cytoplasmatic tail complex and co-localizes with native α_2/β_1 in endothelial cell focal adhesions", *Experimental Cell Research*, 305: 110-121, 2005

[77] **Zhou, A.** „Functional structure of somatomedin B domain of vitronectin", *Protein Science*, 16: 1502-1508, 2007

[78] **Felding-Habermann, B. et al.** „Vitronectin and its receptors", *Current Opinion in Cell Biology*, 5: 864-868, 1993

[79] **Singh, S. et al.** „A NpxY-independent β_5 integrin activation signal regulates phagocytosis of apoptotic cells", *Biochem and Biophys Research Comm*, 364: 540-548, 2007

[80] **Mosher, D. F.** „Adhesive proteins and their cellular receptors", *Cardiovasc Pathol*, Vol. 2, No. 3: 149-155, 1993

[81] **Gahmberg, C. G. et al.** „Regulation of integrin activity and signaling", *General Subjects*, doi:10.1016/j.bbagen.2009.03.007, 2009

[82] **Humphries, M.J.** „Integrin activation: the link between ligand binding and signal transduction", *Current Opinion in Cell Biology*, 8: 632-640, 1996

[83] **Conklin, M. W. et al.** „Local calcium transients contribute to disappearance of pFAK, focal complex removal and deadhesion of neural growth cones and fibroblasts", *Developmental Biology*, 287: 201-212, 2005

[84] **Ginsberg, M. H. et al.** „Inside-out integrin signalling", *Current Opinion in Cell Biology*, Vol. 4: 766-771, 1992

[85] **Lo, S.H.** „Focal adhesions: what's new inside", *Developmental Biology*, 294: 280-291, 2006

[86] **Giancotti, F. G. et al.** „Integrin signaling", *Science*, Vol. 285: 1028-1032, 1999

[87] **Wozniak, M. A. et al.** „Focal adhesion regulation of cell behavior", *Biochimica et Biophysica Acta*, 1692: 103-119, 2004

[88] **Tucker, B. A. et al.** „Src and FAK are key early signalling intermediates required for neutire growth in NGF-responsive adult DRG neurons", *Cellular Signalling*, 20: 241-257, 2008

[89] **Chambard, J.-C. et al.** „ERK implication in cell cycle regulation", *Biochimica et Biophysica Acta*, 1773: 1299-1310, 2007

[90] **Asthagiri, A. R. et al.** „Quantitative relationship among integrin-ligand binding, adhesion, and signaling via focal adhesion kinase and extracellular signal-regulated kinase 2", *Journal of Biological Chemistry*, Vol. 274, No. 38: 27119-27127, 1999

[91] **Schwartz, M. A. et al.** „Integrins and cell proliferation: regulation of cyclin-dependent kinases via cytoplasmatic signaling pathways", *Journal of Cell Science*, 114: 2553-2560, 2001

[92] **Pedersen, S.F. et al.** „The cytoskeleton and cell volume regulation", *Comparative Biochemistry and Physiology Part A*, 130: 385-399, 2001

[93] **Adams, J. C.** „Cell-matrix contact structures", *Cellular Molecular Life* Sciences, 58: 371-392, 2001

[94] **Ingber, D. E.** „Tensegrity 1: cell structure and hierarchical systems biology", *Journal of Cell Science*, 116: 1157-1173, 2003

[95] **Preuss, U.** „Das Aktin-Zytoskelett: ein wichtiger Regulator des Programmierten Zelltods", *BIOspektrum*, 4: 442-446, 2005

[96] **Guillou, H. et al.** „Lamellipodia nucleation by filopodia depends on integrin occupancy and downstream Rac 1 signalling", *Experimental Cell Research*, DOI: 10.1016/j.yexcr.2007.10.026, 2007

[97] **Ngezahayo, A. et al.** „Gap Junction coupling and Apoptosis in GFSHR-17 granulosa cells", *Journal of Membrane Biology*, 204: 137-144, 2005

[98] **Goodenough, D. A. et al.** „Connexins, connexons, and intercellular communication", *Annu Rev Biochem*, 65: 475-50, 1996

[99] **Harris et al.** „Emerging issues of connexin channels: biophysics fill the gap", *Q. Rev. Biophys.*, 34: 325-472, 2001

[100] **Cox, E. A. et al.** „Regulation of integrin-mediated adhesion during cell migration", *Microscopy Research and Technique*, 43: 412-419, 1998

[101] **Imbeault, S. et al.** „The extracellular matrix controls gap junction protein expression and function in postnatal hippocampal neural progenitor cells", *BioMed Central Neuroscience*, DOI :10.1186/1471-2202-10-13, 2009

[102] **Lampe, P. D. et al.** „Cellular interaction of $\alpha_3\beta_1$ with laminin 5 promotes gap junctional communication", *The Journal of Cell Biology*, Vol. 143, No. 6: 1735-1747, 1998

[103] **Plante, I. et al.** „Activation of the integrin linked kinase pathway downregulates hepatic connexin32 via nuclear Akt", *Carcinogenesis*, Vol. 27, No. 9: 1923-1929, 2006

[104] **El-Sabban, M. E. et al.** „ECM-induced gap junctional communication enhances mammary epithelial cell differentiation", *Journal of Cell Science*, Vol. 116: 3531-3541, 2003

[105] **Prochnow, N. et al.** „Connexons and cell adhesion: a romantic phase", *Histochem Cell Biology*, 130: 171-177, 2008

[106] **Koch, L. et al.** „ Towards living tissue substitutes by laser-induced forward transfer of skin cells and human stem cells." *Tissue Engineering Part C*, DOI: 10.1089/ten.tec.2009.0397 (2009)

[107] **Heim, E. et al.** „Fluxgate magnetorelaxometry of superparagmetic nanoparticles for hydrogel characterization", *Journal of Magnetism and Magnetic Materials*, 311: 150-154, 2007

[108] **Begandt, D. et al.** „Dipyridamole increases gap junction coupling in bovine GM-7373 aortic endothelial cells by a cAMP-protein kinase A dependent pathway." *British Journal of Pharmacology*, DOI 10.1007/s10863-009-9262-2 (2009)

[109] **Bacso, Z. et al.** „Measurement of DNA damage associated with apoptosis by laser scanning cytometry", *Cytometry*, 45: 180-186, 2001

[110] **Jeon, O. et al.** „Mechanical properties and degradation behaviors of hyaluronic acid hydrogel cross-linked at various cross-linking densities", *Carbohydrate Polymers*, 2007

[111] **Arima, Y. et al.** „Effect of wettability and surface functional groups on protein adsorption and cell adhesion using well-defined mixed self-assembled monolayers", *Biomaterials*, 28: 3074-3082, 2007

[112] **Tzoneva, R. et al.** „Wettability of substrata controls cell-substrate and cell-cell adhesions", *Biochimica et Biophysica Acta*, 1770: 1538-1547, 2007

[113] **Velzenberger, E. et al.** „Characterization of biomaterials polar interactions in physiological conditions using liquid-liquid contact angle measurements relation to fibronectin adsorption", *Colloids and Surfaces B: Biointerfaces*, 68: 238-244, 2009

[114] **Bryant, S. J. et al.** „Manipulations in hydrogel chemistry control photoencapsulated chondrocyte behavior and their extracellular matrix production", *Journal Biomed Mater Res*, 67 A: 1430-1436, 2003

[115] **Yu, L. M. Y. et al.** „Promoting neuron adhesion and growth", *Materials Today*, 11: 36-43, 2008

[116] **Norman, L. L. et al.** „Guiding axons in the central nervous system: a tissue engineering approach", *Tissue Engineering*, 15: DOI 10.1089, 2009

[117] **Nicolas, A. et al.** „Limitation of cell adhesion by the elasticity of the extracellular matrix", *Biophysical Journal*, 91: 61-73, 2006

[118] **Ingber, D. E.** „Tensegrity 2: how structural networks influence cellular information processing networks", *Journal of Cell Science*, 116: 1397-1408, 2003

[119] **Tan, J. et al.** „Biomaterials with hierarchically defined micro-and nanoscale structure", *Biomaterials*, 25: 3593-3601, 2004

[120] **Andersson, A. S. et al.** „Nanoscale features influence epithelial cell morphology and cytokine production", *Biomaterials*, 24: 3427-3436, 2003

[121] **Dalby, M. J. et al.** „Use of nanotopography to study mechanotransduction in fibroblasts – methods and perspectives", *European Journal of Cell Biology*, 83: 159-169, 2004

[122] **Zinger, O. et al.** „Time-dependent morphology and adhesion of osteoblastic cells on titanium model surfaces featuring scale-resolved topography", *Biomaterials*, 25: 2695-2711, 2004

[123] **Carman, M. L. et al.** „Engineered antifouling microtopographies – correlating wettability with cell attachment", *Biofouling*, 1-11, 2006

[124] **Berry, C. C. et al.** „The fibroblast response to tubes exhibiting internal nanotopography", *Biomaterials*, 26: 4985-4992, 2005

[125] **Keselowsky, B. G. et al.** „Quantitative methods for analysis of integrin binding and focal adhesion formation on biomaterial surfaces", *Biomaterials*, 26: 413-418, 2005

[126] **Auernheimer, J. et al.** „Photoswitched cell adhesion on surfaces with RGD peptides", *J. Am. Chem. Soc.*, 127: 16107-16110, 2005

[127] **Kurihara, H. et al.** „Cell adhesion ability of artificial extracellular matrix proteins containing a long repetitive Arg-Gly-Asp sequence", *Journal of Bioscience and Bioengineering*, Vol. 100, No. 1: 82-87, 2005

[128] **Mosher, D. F.** „Physiology of fibronectin", *Annual Reviews*, 35: 561-575, 1984

[129] **Candé, C. et al.** „Apoptosis inducing factor (AIF): key to the conserved caspase-independent pathway of cell death?", *Journal of Cell Science*, 115: 4727-4737, 2002

[130] **Lynch, K. et al.** „Basic fibroblast growth factor inhibits apoptosis of spontaneously immortalized granulosa cells by regulating free intracellular Ca^{2+} levels through a proteinkinase C δ-dependent pathway", *Endocrinology*, Vol. 141, No. 11: 4209-4217, 2000

[131] **Buttiglione, M. et al.** „Behaviour of SH-SY5Y neuroblastoma cell line grown in different media and on different chemical modified substrates", *Biomaterials*, 28: 2932-2945, 2007

[132] **Bongrand, P. et al.** „Physics of cell adhesion", *Progress in Surface Science*, 12: 217-286, 1982

[133] **García, A. J. et al.** „Modulation of cell proliferation and differentiation through substrate-dependent changes in fibronectin conformation", *Molecular and Cellular Biology*, Vol. 10, No. 3: 785-798, 1999

[134] **Velzenberger, E. et al.** „Characterization of biomaterials polar interactions in physiological conditions using liquid-liquid contact angle measurements relation to fibronectin adsorption", *Colloids and Surfaces B: Biointerfaces*, 68: 238-244, 2009

[135] **Meadows, P. Y. et al.** „Force microscopy study of fibronectin adsorption and subsequent cellular adhesion to substrates with well-defined surface chemistries", *Langmuir*, 21: 4096-4107, 2005

[136] **Zhang, Y. et al.** „Angiopoietin-related growth factor (AGF) supports adhesion, spreading, and migration of keratinocytes, fibroblasts, and endothelial cells through interaction with RGD-binding integrins", *Biochem and Biophys Research Comm*, 347: 100-108, 2006

[137] **Elliott, J. T. et al.** „The effect of surface chemistry on the formation of thin films of native fibrillar collagen", *Biomaterials*, 28: 576-585, 2007

[138] **Kingsley, K. et al.** „ERK 1 / 2 mediates PDGF-BB stimulated vascular smooth muscle cell proliferation and migration on laminin-5", *Biochem and Biophys Research Comm*, 293: 1000-1006, 2002

[139] **Grinnell, F. et al.** „Fibronectin adsorption on hydrophilic and hydrophobic surfaces detected by antibody binding and analysed during cell adhesion in serum-containing medium", *Journal of Biological Chemistry*, Vol. 257, No. 9: 4888-4893, 1982

[140] **Shanker, A. J. et al.** „Matrix protein-specific regulation of Cx43 expression in cardiac myocytes subjected to mechanical load", *Circulation Research*, 96: 559-566, 2005

[141] **Parker, K. K. et al.** „Extracellular matrix, mechanotransduction and structural hierachies in heart tissue engineering", *Phil. Tran. R. Soc. B*, 362: 1267-1279, 2007

[142] **Saffitz, J. E. et al.** „Effects of mechanical forces and mediators of hypertrophy on remodeling of gap junctions in the heart", *Circulation Research*, 94: 585-591, 2004

[143] **Mineur, P. et al.** „RGDS and DGEA-induced $[Ca^{2+}]_i$ signalling in human dermal fibroblasts", *Biochimica et Biophysica Acta*, 1746: 28-37, 2005

[144] **Faucheux, N. et al.** „The dependence of fibrillar adhesions in human fibroblasts on substratum chemistry",*Biomaterials*, 27: 234-245, 2006

[145] **Ivaska, J. et al.** „Novel functions of vimentin in cell adhesion, migration, and signaling", *Experimental Cell Research*, 313: 2050-2062, 2007

[146] **Kim, S.-Y. et al.** „RGD-peptide presents anti-adhesive effect, but no direct pro-apoptotic effect on endothelial progenitor cells", *Archives of Biochem and Biophys*, 459: 40-49, 2007

[147] **Zhang, Y. et al.** „Angiopoietin-related growth factor (AGF) supports adhesion, spreading, and migration of keratinocytes, fibroblasts, and endothelial cells through interaction with RGD-binding integrins", *Biochem and Biophys Research Comm*, 347: 100-108, 2006

[148] **Mawatari, K. et al.** „Activation of integrin receptors is required for growth factor – induced smooth muscle cell dysfunction", *J Vasc Surg*, 31: 375-381, 2000

[149] **Allen, L. T. et al.** „Surface-induced changes in protein adsorption and implictions for cellular phenotype responses to surface interaction", *Biomaterials*, 27: 3096-3108, 2006

[150] **Groth, T. et al.** „Studies on cell-biomaterial interaction: role of tyrosine phosphorylation during fibroblast spreading on surfaces varying in wettability", *Biomaterials*, 17: 1227-1234, 1996

[151] **Baujard-Lamotte, L. et al.** „Kinetics of conformational changes of fibronectin adsorbed onto model surfaces", *Colloids and Surfaces B: Biointerfaces*, 63: 129-137, 2008

[152] **Keselowsky, B. G. et al.** „Surface chemistry modulates focal adhesion composition and signaling through changes in integrin binding", *Biomaterials*, 25: 5947-5954, 2004

[153] **Pereira, P. et al.** „Interdomain mobility and conformational stability of type III fibronectin domain pairs control surface adsorption, desorption and unfolding", *Colloids and Surfaces B: Biointerfaces*, 64: 1-9, 2008

[154] **McFarland, C. D. et al.** „Attachment of cultured human bone cells to novel polymers", *Journal Biomed Mater Res*, 44: 1-11, 1999

[155] **Khang, D. et al.** „Enhanced fibronectin adsorption on carbon nanotube/poly(carbonate) urethane: independent role of surface nano-roughness and associated surface energy", *Biomaterials*, 28: 4756-4768, 2007

[156] **Martínez, E. et al.** „Focused ion beam/scanning electron microscopy characterization of cell behavior on polymer micro- /nanopatterned substrates: a study of cell-substrate interactions", *Micron*, 2007

7.2 Figures

Figure 1: Two-photon polymerization 9
Figure 2: Schematic image of laser printing setup 10
Figure 3: Extracellular matrix including adhesion ligands and receptors 14
Figure 4: Schematic image of integrin receptors 17
Figure 5: Schematic image of outside-in signaling in adhesion mechanism 18
Figure 6: Model of (a) FAK and (b) Shc pathways [88] 19
Figure 7: Schematic image of gap junctions 22
Figure 8: Shaking table for three-dimensional structures 30
Figure 9: Comet assay parameters according to *autocomet.com* 32
Figure 10: Formulas for calculating the kinetic of adhesion mechanism 33
Figure 11: Quantification of cell morphology 35
Figure 12: Quantification of gap junction coupling showing rectangles (250 x 100 px) and plot profiles of (a) diffusion distance of lucifer yellow and (b) background 37
Figure 13: Proliferation profiles of (a) GFSHR-17 granulosa cells, (b) GM-7373 endothelial cells, and (c) SH-SY5Y neuroblastoma cells, (d) doubling times [h] on polymer Ormocomp® in comparison to the control over 48 h or 96 h cultivation time ... 41
Figure 14: Proliferation profiles of (a) human fibroblasts, (b) GM-7373 endothelial cells, and (c) SH-SY5Y neuroblastoma cells on polymer silicone elastomer in comparison to the control over 48 h cultivation time. 42
Figure 15: Proliferation profiles of (a) human fibroblasts, (b) GM-7373 endothelial cells, and (c) SH-SY5Y neuroblastoma cells on hydrogel HESHEMA in dependence of the DS-value (0.07, 0.11, 0.2) in comparison to the control over 48 h cultivation time. 43
Figure 16: Proliferation profiles of (a) human fibroblasts, (b) GM-7373 endothelial cells, and (c) SH-SY5Y neuroblastoma cells on hydrogel PEG SR610 (2 wt% photoinitiator Irgacure 2959) in dependence of material aging in comparison to the control over 48 h cultivation time. 44
Figure 17: SEM images of laser-fabricated three-dimensional scaffolds composed of PEG SR610 45
Figure 18: Microscopic images of human fibroblasts cultivated on laser-fabricated grating structures of Ormocomp® in dependence of grating size. 46
Figure 19: Stereo microscope images of SH-SY5Y neuroblastoma cells cultivated on laser-fabricated cylinders composed of Ormocomp® 46
Figure 20: Microscopic images of GM-7373 endothelial cells on laser-fabricated PEG SR610 scaffolds. 47
Figure 21: Microscopic images of NIH3T3 fibroblasts on laser-fabricated PEG SR610 scaffolds with varying diameters of (a) 50 μm and (b) 70 μm after 4 days cultivation time. 48
Figure 22: Fluorescence image of NIH3T3 fibroblasts (bright gray) and HaCaT keratinocytes (dark gray) arranged by laser-induced forward transfer. 48

Figure 23: Proliferation profiles of (a) NIH3T3 fibroblasts and (b) HaCaT keratinocytes after laser-induced forward transfer (LIFT) and under control conditions over 96 h cultivation time. 50

Figure 24: SEM images of laser-fabricated spike structures in silicon at different laser fluences [J/cm²]. 50

Figure 25: SEM images of (a) laser-fabricated silicon spikes and (b) the negative replicas in silicone elastomer. 51

Figure 26: SEM images of hierarchical nano- and microsuperimposed surface structures in titanium with different magnifications. 51

Figure 27: Histograms of groove structures in titanium in dependence of groove width. (a) 5 µm, (b) 10 µm, (c) 15 µm, (d) 20 µm, (e) 25 µm, and (f) 30 µm. 52

Figure 28: SEM images of laser-fabricated nanostructures in platinum at different laser processing parameters. (a - c) nanoroughness and (d - e) nanogrooves. 53

Figure 29: Fluorescence images (stained actin filaments) of human fibroblasts (a - f) and MG-63 osteoblasts (g - l) cultivated on groove structures in titanium in dependence of groove width after 24 h cultivation time. 56

Figure 30: Fluorescence images of human fibroblasts on (a) control and (b) hierarchical titanium structures, and of MG-63 osteoblasts on (c) control and (d) hierarchical titanium structures after 24 h cultivation time. 57

Figure 31: Quantification of cell morphology of human fibroblasts and SH-SY5Y neuroblastoma cells cultivated on silicon and silicon spikes in comparison to the control after 24 h cultivation time. 58

Figure 32: Proliferation profiles of (a) human fibroblasts and (b) MG-63 osteoblasts cultivated on hierarchical nano- and micro- superimposed titanium structures and under control conditions over 48 and 72 h cultivation time. 59

Figure 33: Proliferation profiles of human fibroblasts cultivated on nanostructured platinum and under control conditions for 48 h cultivation time. 60

Figure 34: Control adhesion pattern via SRIC-technique of (a) human fibroblasts, (b) GM-7373 endothelial cells, (c) SH-SY5Y neuroblastoma cells, (d) HaCaT keratinocytes, (e) MG-63 osteoblasts, and (f) A10 smooth muscle cells after 24 h cultivation time. 62

Figure 35: Histogram of adhesion pattern of (a) SH-SY5Y neuroblastoma cells and (b) MG-63 osteoblasts on adhesion ligands after 5 h cultivation time. 63

Figure 36: Fluorescence images of SH-SY5Y neuroblastoma cells on (a) control, (b) laminin, (c) fibronectin, (d) collagen, and (e) vitronectin after 5 h cultivation time. 66

Figure 37: Quantification of cell morphology of human fibroblasts cultivated on adhesion ligands in comparison to the control after 5 h cultivation time. 66

Figure 38: Quantification of cell morphology of GM-7373 endothelial cells cultivated on adhesion ligands in comparison to the control after 5 h cultivation time. 67

Figure 39: Quantification of cell morphology of SH-SY5Y neuroblastoma cells cultivated on adhesion ligands in comparison to the control after 5 h cultivation time. ... 68

Figure 40: Quantification of cell morphology of HaCaT keratinocytes cultivated on adhesion ligands in comparison to the control after 5 h cultivation time. 68

Figure 41: Quantification of cell morphology of MG-63 osteoblasts cultivated on adhesion ligands in comparison to the control after 5 h cultivation time. 69

Figure 42: Quantification of cell morphology of A10 smooth muscle cells cultivated on adhesion ligands in comparison to the control after 5 h cultivation time. 69

Figure 43: Adhesion profiles of (a) human fibroblasts, (b) GM-7373 endothelial cells, and (c) SH-SY5Y neuroblastoma cells on adhesion ligands over 5 h cultivation time... 70

Figure 44: Quantification of cell morphology of human fibroblasts cultivated on adhesion ligands in comparison to the control over 24 h cultivation time. 75

Figure 45: Quantification of cell morphology of GM-7373 endothelial cells cultivated on adhesion ligands in comparison to the control over 24 h cultivation time. 76

Figure 46: Quantification of cell morphology of SH-SY5Y neuroblastoma cells cultivated on adhesion ligands in comparison to the control over 24 h cultivation time. 77

Figure 47: Quantification of cell morphology of HaCaT keratinocytes cultivated on adhesion ligands in comparison to the control over 24 h cultivation time. 77

Figure 48: Quantification of cell morphology of MG-63 osteoblasts cultivated on adhesion ligands in comparison to the control over 24 h cultivation time. 78

Figure 49: Quantification of cell morphology of A10 smooth muscle cells cultivated on adhesion ligands in comparison to the control over 24 h cultivation time. 79

Figure 50: Proliferation profiles of (a) human fibroblasts, (b) GM-7373 endothelial cells, (c) SH-SY5Y neuroblastoma cells, (d) HaCaT keratinocytes, (e) MG-63 osteoblasts, and (f) A10 smooth muscle cells in dependence of adhesion ligands in comparison to the control. 81

Figure 51: Fluorescence images of cell to cell communication over gap junction channels visualized with lucifer yellow after scrape loading procedure. 84

7.3 Tables

Table 1: Analysis of DNA damage effects of PEG SR610 (2 wt% photoinitiator Irgacure 2959) in dependence of material aging demostrated by comet assay of human fibroblasts, GM-7373 endothelial cells, and SH-SY5Y neuroblastoma cells. 39

Table 2: Adhesion time A_T [h] of human fibroblasts, GM-7373 endothelial cells, and SH-SY5Y neuroblastoma cells on silicone elastomer, HESHEMA (DS 0.11), and aged PEG SR610 (2 wt% photoinitiator Irgacure 2959). 40

Table 3: Analysis of DNA damage effects after laser-induced forward transfer of NIH3T3 fibroblasts, HaCaT keratinocytes, human mesenchymal stem cells, porcine mesenchymal stem cells in comparison to the control demonstrated by comet assay.. 49

Table 4: Water contact angle measurements of silicon, silicone elastomer, titanium, and platinum in dependence of surface structuring. 53

Table 5: Analysis of DNA damage effects of surface topographies in silicon, silicone elastomer, titanium, and platinum demonstrated by comet assay of human fibroblasts and SH-SY5Y neuroblastoma cells in comparison to controls after 24 h cultivation time. 54

Table 6: Quantification of cell orientation of human fibroblasts and MG-63 osteoblasts on groove structures of titanium in dependence of groove width after 24 h cultivation time. 55

Table 7: Proliferation results of human fibroblasts and SH-SY5Y neuroblastoma cells cultivated on unstructured and structured silicon and silicone elastomer in comparison to the control after 48 h. 59

Table 8: Adhesion time A_T [h] of human fibroblasts, GM-7373 endothelial cells, SH-SY5Y neuroblastoma cells, HaCaT keratinocytes, MG-63 osteoblasts, and A10 smooth muscle cells under control conditions. 61

Table 9: Quantification of adhesion pattern of human fibroblasts, GM-7373 endothelial cells, SH-SY5Y neuroblastoma cells, HaCaT keratinocytes, MG-63 osteoblasts, and A10 smooth muscle cells in dependence of adhesion ligands after 5 h cultivation time. 64

Table 10: Adhesion time A_T [min] of human fibroblasts, GM-7373 endothelial cells, SH-SY5Y neuroblastoma cells, HaCaT keratinocytes, MG-63 osteoblasts, and A10 smooth muscle cells on adhesion ligands in dependence of the ligand concentration.. 71

Table 11: Quantification of adhesion pattern of human fibroblasts, GM-7373 endothelial cells, SH-SY5Y neuroblastoma cells, HaCaT keratinocytes, MG-63 osteoblasts, and A10 smooth muscle cells in dependence of adhesion ligands after 24 h cultivation time. 73

Table 12: Doubling time [h] of human fibroblasts, GM-7373 endothelial cells, SH-SY5Y neuroblastoma cells, HaCaT keratinocytes, MG-63 osteoblasts, and A10 smooth muscle cells on adhesion ligands in dependence of the ligand concentration.. 82

Table 13: Quantification of gap junction coupling of human fibroblasts, GM-7373 endothelial cells, HaCaT keratinocytes, and A10 smooth muscle cells on adhesion ligands in dependence of the ligand concentration. 85

7.4 Abbrevations

AMIDAS	adjacent to MIDAS	HF	hydrofluorid acid
A_T	adhesion time	ILK	integrin linked kinase
Bis	4-bis diethylaminobenzophenone	IRG	irgacure
CDK	cyclin dependent kinase	JNK	c-Jun amino-terminal kinase
DAPI	4', 6-diamidino-2-phenylindole-dihydrochlorid:hydrat	LIFT	laser-induced forward transfer
DMEM	Dulbeccoe's Modified Eagles Medium	LIMBS	ligand induced metal binding site
DS	degree of substitution	MAPK	mitogen-activated protein kinase
EDTA	ethylen-diamin-tetra-acetat	MIDAS	metal ion dependent adhesion site
EDX	energy dispersive X-ray spectroscopy	PEG	poly(ethylene)glycol diacrylate
ERK	extracellular signal-regulated kinase	PBS	phosphate buffer salt
FAK	focal adhesion kinase	PI 3K	phosphoinoside 3-kinase
FCS	fetal calf serum	RGD	Arg-Gly-Asp binding sequence
HEMA	hydroxyethylmethacrylate	SEM	standard error of mean
HES	hydroxyethylstarch	SEM	scanning electron microscopy
HESHEMA	hydroxymethacrylat-hyroxyethylstarch	SRIC	surface reflectance interference contrast

7.5 Software

Word processing	Microsoft XP Professional Word
Statistical evaluation	Microsoft XP Professional Excel, Origin 7.0
Image processing	Comet score (http://autocomet.com), Image J (http://rsbweb.nih.gov/ij/), Adope Photoshop 7.0
Camera software	Xaw TV, NIS Elements AR 3.0 and E Z-C1 3.5 (Nikon, Düsseldorf, Germany)

7.6 Media and solutions

7.6.1 Ligand coating

Ligands (Sigma-Aldrich, Taufkirchen, Germany)	Stocking solutions [wt %]	Coating concentration [µg/cm²]
Laminin (sarcoma basement	0.01 in PBS	2, 1
Fibronectin (bovine plasma)	0.01 in PBS	5, 3, 1
Collagen type I (rat tail)	0.01 in H$_2$O	10, 8, 6
Vitronectin (bovine plasma)	1 ml H$_2$O	0.1

7.6.2 Cell culture

	Composition
DMEM	**Dulbecco's Modified Eagle Medium** (D8900) (Sigma-Aldrich, Taufkirchen, Germany)
	Penicilin (100 U/ml), Streptomycin (100 µg/ml), Patricin (0.5 µg/ml) (Gibco, Carlsbad, USA)
	5 % or 10 % fetal calf serum
	pH 7.4; 300 ± 5 mosmol
PBS	**Phosphate Buffered Salt [mM]**
	137 NaCl, 2.7 KCl, 9.86 Na$_2$HPO$_4$, 1.14 KH$_2$PO$_4$
	pH 7.4
PBS + EDTA	**Phosphate Buffered Salt + Ethylene-Diamine-Tetra-Acetate [mM]**
	PBS + 3.4 EDTA
	pH 7.4
Trypsin	0.25 % in PBS + EDTA
	pH 7.4

7.6.3 Analysis of DNA damage effects

	Composition
Agarose	0.6 % low melting agarose in PBS
Lysis buffer	2.5 M NaCl, 100 mM Na_2EDTA, 10 mM Tris, 1 % lauryl sarcosin, 1 % Triton X-100, 10 % DMSO pH 10
Electrophoresis buffer	1 mM Na_2EDTA, 300 mM NaOH pH > 13
Staining	20 µg/ml ethidium bromide
Neutralization	400 mM Tris pH 7.4

7.6.4 Staining solutions

	Composition
Fixation	4 % formaldehyde in PBS
Permeabilization	0.3 % Triton X-100 in PBS
Nucleus staining	1 µM DAPI (4', 6-Diamidino-2Phenylindole-Dihydrochloride:Hydrate) in PBS (Molecular probes Invitrogen, Grenzach-Whylen, Germany)
Actin filaments staining	0.6 U phalloidin-Alexa 488 in PBS (Molecular probes Invitrogen, Grenzach-Whylen, Germany)
Conservation	PBS

7.6.5 Gap junction coupling

	Composition
Washing solution	NaCl-BS [mM] 121 NaCl, 5.4 KCl, 25 HEPES, 0.8 $MgCl_2*6H_2O$, 5.5 Glucose, 6 $NaHCO_3$, 1.8 $CaCl_2*2H_2O$ pH 7.4; 295 ± 5 mosmol
Staining solution	0.25 % Lucifer Yellow in NaCl-BS (Sigma-Aldrich, Taufkirchen, Germany)
Fixation	4 % formaldehyde in PBS
Conservation	PBS

7.7 Analysis of cell morphology – cell sizes [μm]

Biomaterial-cell interaction were also characterized via analysis of cell morphology. Besides counting the average number of extensions such as filopodia, lamellipodia and retraction fibres which were defined as appendages that taper off to the surface and to neighboring cells, nucleus and cell dilations were calculated. These dilations were given as the ratio of length and width (L_n/W_n and L_c/W_c, respectively). Length and width of the nuclei and cells were recorded with the help of ImageJ software and given in micrometer scale. In the following both paramters are shown as average ± SEM for each experiment. At leat 100 cells per treatment were evaluated.

Cell sizes of human fibroblasts and SH-SY5Y neuroblastoma cells on silicon spike structures in comparison to the control after 24 h cultivation time (Figure 31)

		Cell size [μm]			
		Nucleus		Cell	
Cell type	Treatment	Length (L_n)	Width (W_n)	Length (L_c)	Width (W_c)
Fibroblasts	Control	17.32 ± 0.47	11.25 ± 0.34	124.29 ± 6.36	20.49 ± 0.96
	Silicon	16.12 ± 0.61	10.04 ± 0.42	75.85 ± 3.89	20.99 ± 1.1
	Silicon spikes	11.2 ± 0.44	6.05 ± 0.21	32.65 ± 2.35	12.44 ± 0.51
Neuroblastoma	Control	15.92 ± 0.37	7.35 ± 0.23	68.6 ± 3.03	16.29 ± 1.24
	Silicon	15.33 ± 0.35	9.23 ± 0.36	70.07 ± 2.99	18.07 ± 1.31
	Silicon spikes	13.91 ± 0.27	9.79 ± 0.22	39.03 ± 2.57	14.32 ± 0.32

Cell sizes of human fibroblasts in dependence of adhesion ligands after 5 h cultivation time (Figure 37)

	Cell size [μm]			
	Nucleus		Cell	
Treatment	Length (L_n)	Width (W_n)	Length (L_c)	Width (W_c)
Control	19.19 ± 0.87	11.81 ± 0.63	121.56 ± 8.09	44.33 ± 5.67
Laminin 2 μg/cm²	19.95 ± 0.54	11.73 ± 0.27	98.94 ± 7.08	32.59 ± 2.75
Fibronectin 5 μg/cm²	19.56 ± 0.71	11.56 ± 0.54	110.62 ± 7.15	33.11 ± 2.52
Collagen 10 μg/cm²	19.95 ± 0.38	11.97 ± 0.48	106.33 ± 5.37	36.89 ± 3.01
Vitronectin 0.1μg/cm²	20.11 ± 0.46	11.84 ± 0.47	119.03 ± 4.59	42.52 ± 4.82

Cell sizes of GM-7373 endothelial cells in dependence of adhesion ligands after 5 h cultivation time (Figure 38)

Treatment	Cell size [µm]			
	Nucleus		Cell	
	Length (L_n)	Width (W_n)	Length (L_c)	Width (W_c)
Control	13.55 ± 0.31	9.25 ± 0.33	67.15 ± 2.81	24.78 ± 1.29
Laminin 2 µg/cm²	13.7 ± 0.45	9.88 ± 0.38	51.94 ± 2.86	24.09 ± 1.4
Fibronectin 5 µg/cm²	12.73 ± 0.35	10.57 ± 1.43	46.28 ± 2.97	20.39 ± 1.21
Collagen 10 µg/cm²	12.89 ± 0.47	8.89 ± 0.34	47.64 ± 2.91	23.07 ± 1.29
Vitronectin 0.1µg/cm²	14.41 ± 0.38	9.92 ± 0.38	59.21 ± 3.11	34.33 ± 1.55

Cell sizes of SH-SY5Y neuroblastoma cells in dependence of adhesion ligands after 5 h cultivation time (Figure 39)

Treatment	Cell size [µm]			
	Nucleus		Cell	
	Length (L_n)	Width (W_n)	Length (L_c)	Width (W_c)
Control	13.82 ± 0.58	9.97 ± 0.48	30.29 ± 2.25	17.62 ± 0.74
Laminin 2 µg/cm²	13.34 ± 0.63	9.03 ± 0.57	53.93 ± 3.41	14.68 ± 0.84
Fibronectin 5 µg/cm²	12.04 ± 0.49	8.67 ± 0.39	25.03 ± 1.8	14.8 ± 0.63
Collagen 10 µg/cm²	14.23 ± 0.44	10.56 ± 0.58	25.57 ± 1.74	17.46 ± 0.64
Vitronectin 0.1µg/cm²	12.75 ± 0.44	8.55 ± 0.28	53.92 ± 3.8	15.77 ± 1.76

Cell sizes of HaCaT keratinocytes in dependence of adhesion ligands after 5 h cultivation time (Figure 40)

Treatment	Cell size [µm]			
	Nucleus		Cell	
	Length (L_n)	Width (W_n)	Length (L_c)	Width (W_c)
Control	12.21 ± 0.41	9.02 ± 0.29	27.47 ± 2.1	18.42 ± 1.22
Laminin 2 µg/cm²	13.94 ± 0.61	9.49 ± 0.38	33.62 ± 1.79	20.19 ± 1.76
Fibronectin 5 µg/cm²	14.05 ± 0.54	9.33 ± 0.44	48.56 ± 3.15	18.06 ± 1.73
Collagen 10 µg/cm²	14.46 ± 0.65	8.15 ± 0.34	55.16 ± 3.39	15.84 ± 0.7
Vitronectin 0.1µg/cm²	13.71 ± 0.62	9.22 ± 0.43	30.85 ± 1.83	18.21 ± 1.48

Cell sizes of MG-63 osteoblasts in dependence of adhesion ligands after 5 h cultivation time (Figure 41)

Treatment	Cell size [µm]			
	Nucleus		Cell	
	Length (L_n)	Width (W_n)	Length (L_c)	Width (W_c)
Control	20.62 ± 0.69	12.79 ± 0.58	66.68 ± 5.37	21.37 ± 1.2
Laminin 2 µg/cm²	20.83 ± 0.61	12.12 ± 0.61	72.92 ± 4.35	19.85 ± 1.55
Fibronectin 5 µg/cm²	20.24 ± 0.64	13.82 ± 0.59	68.27 ± 4.31	22.14 ± 1.31
Collagen 10 µg/cm²	22.56 ± 0.57	12.69 ± 0.68	90.52 ± 4.97	20.18 ± 1.83
Vitronectin 0.1µg/cm²	21.47 ± 0.49	12.22 ± 0.59	80.54 ± 4.2	18.8 ± 1.34

Cell sizes of A10 smooth muscle cells in dependence of adhesion ligands after 5 h cultivation time (Figure 42)

Treatment	Cell size [μm]			
	Nucleus		Cell	
	Length (L_n)	Width (W_n)	Length (L_c)	Width (W_c)
Control	24.65 ± 0.82	14.86 ± 0.46	68.21 ± 3.48	42.24 ± 2.13
Laminin 2 μg/cm²	25.76 ± 0.67	16.15 ± 0.42	81.56 ± 3.23	39.36 ± 4.11
Fibronectin 5 μg/cm²	23.86 ± 0.85	15.09 ± 0.65	63.61 ± 4.11	35.93 ± 1.42
Collagen 10 μg/cm²	25.41 ± 0.76	17.32 ± 0.43	69.54 ± 3.71	42.07 ± 2.13
Vitronectin 0.1μg/cm²	24.76 ± 0.62	16.15 ± 0.46	66.29 ± 2.57	44.24 ± 1.6

Cell sizes of human fibroblasts in dependence of adhesion ligands after 24 h cultivation time (Figure 44)

Treatment	Cell size [μm]			
	Nucleus		Cell	
	Length (L_n)	Width (W_n)	Length (L_c)	Width (W_c)
Control	17.32 ± 0.47	11.25 ± 0.34	124.29 ± 6.36	20.49 ± 0.96
Laminin 2 μg/cm²	18.34 ± 0.42	12.39 ± 0.4	114.4 ± 5.07	35.48 ± 2.45
Laminin 1 μg/cm²	18.64 ± 0.46	12.88 ± 0.39	109.86 ± 4.97	34.78 ± 1.75
Fibronectin 5 μg/cm²	18.76 ± 0.51	12.11 ± 0.37	116.6 ± 6.68	30.31 ± 1.96
Fibronectin 3 μg/cm²	19.19 ± 0.44	12.51 ± 0.36	119.56 ± 4.31	41.29 ± 3.26
Fibronectin 1 μg/cm²	18.77 ± 0.49	12.18 ± 0.34	113.83 ± 5.5	32.05 ± 2.21
Collagen 10 μg/cm²	19.35 ± 0.49	12.59 ± 0.44	122.99 ± 5.25	31.39 ± 2.35
Collagen 8 μg/cm²	20.42 ± 0.36	12.81 ± 0.64	114.77 ± 5.31	35.29 ± 2.25
Collagen 6 μg/cm²	20.71 ± 0.48	14.08 ± 0.54	129.08 ± 6.19	36.51 ± 1.95
Vitronectin 0.1μg/cm²	17.92 ± 0.62	11.48 ± 0.49	89.94 ± 5.57	28.24 ± 1.99

Cell sizes of GM-7373 endothelial cells in dependence of adhesion ligands after 24 h cultivation time (Figure 45)

Treatment	Cell size [μm]			
	Nucleus		Cell	
	Length (L_n)	Width (W_n)	Length (L_c)	Width (W_c)
Control	14.53 ± 0.19	10.71 ± 0.2	95.99 ± 2.1	24.25 ± 0.92
Laminin 2 μg/cm²	15.34 ± 0.39	12.47 ± 0.36	93.41 ± 4.2	33.16 ± 1.47
Laminin 1 μg/cm²	15 ± 0.35	11.29 ± 0.24	85.39 ± 3.64	33.74 ± 1.34
Fibronectin 5 μg/cm²	13.69 ± 0.4	9.99 ± 0.26	30.99 ± 2.37	17.26 ± 1
Fibronectin 3 μg/cm²	14.54 ± 0.4	11.56 ± 0.27	71.76 ± 2.65	36.57 ± 1.42
Fibronectin 1 μg/cm²	14.58 ± 0.35	11.7 ± 0.31	76.45 ± 3.18	32.49 ± 1.3
Collagen 10 μg/cm²	14.17 ± 0.37	10.58 ± 0.38	56.41 ± 4.46	24.67 ± 1.38
Collagen 8 μg/cm²	14.89 ± 0.26	11.89 ± 0.29	72.72 ± 2.79	39.25 ± 1.61
Collagen 6 μg/cm²	14.64 ± 0.27	12.05 ± 0.29	70.45 ± 2.99	36.09 ± 1.25
Vitronectin 0.1μg/cm²	16.12 ± 0.28	13.28 ± 0.3	86.61 ± 3.2	36.62 ± 1.4

Cell sizes of SH-SY5Y neuroblastoma cells in dependence of adhesion ligands after 24 h cultivation time (Figure 46)

Treatment	Cell size [µm]			
	Nucleus		Cell	
	Length (L_n)	Width (W_n)	Length (L_c)	Width (W_c)
Control	15.92 ± 0.37	7.35 ± 0.23	68.6 ± 3.03	16.28 ± 1.24
Laminin 2 µg/cm²	14.45 ± 0.44	9.92 ± 0.38	55.23 ± 2.81	17.55 ± 0.73
Laminin 1 µg/cm²	12.56 ± 0.32	7.66 ± 0.31	40.25 ± 2.23	13.63 ± 0.52
Fibronectin 5 µg/cm²	14.64 ± 0.41	10.11 ± 0.35	34.47 ± 2.03	18.88 ± 0.72
Fibronectin 3 µg/cm²	11.53 ± 0.36	7.84 ± 0.29	26.84 ± 1.51	14.1 ± 0.56
Fibronectin 1 µg/cm²	13.52 ± 0.44	8.69 ± 0.34	35.36 ± 2.06	15.53 ± 0.73
Collagen 10 µg/cm²	14.17 ± 0.32	10.68 ± 0.34	42.03 ± 2.47	18.77 ± 0.74
Collagen 8 µg/cm²	14.19 ± 0.39	9.23 ± 0.31	42.72 ± 2.36	16.12 ± 0.64
Collagen 6 µg/cm²	12.75 ± 0.39	8.67 ± 0.31	49.22 ± 2.65	16.92 ± 0.83
Vitronectin 0.1µg/cm²	16.52 ± 0.42	10.1 ± 0.37	68.99 ± 3.99	18.43 ± 0.81

Cell sizes of HaCaT keratinocytes in dependence of adhesion ligands after 24 h cultivation time (Figure 47)

Treatment	Cell size [µm]			
	Nucleus		Cell	
	Length (L_n)	Width (W_n)	Length (L_c)	Width (W_c)
Control	16.28 ± 0.33	11.64 ± 0.35	46.83 ± 1.59	23.16 ± 1.2
Laminin 2 µg/cm²	16.54 ± 0.5	11.54 ± 0.41	46.29 ± 2.19	22.5 ± 0.98
Laminin 1 µg/cm²	15.86 ± 0.42	10.94 ± 0.42	39.89 ± 1.95	20.84 ± 0.81
Fibronectin 5 µg/cm²	15.68 ± 0.47	10.76 ± 0.32	42.78 ± 1.65	21.23 ± 0.79
Fibronectin 3 µg/cm²	15.78 ± 0.51	10.6 ± 0.4	40.55 ± 1.71	20.84 ± 0.92
Fibronectin 1 µg/cm²	15.88 ± 0.38	10.94 ± 0.36	41.72 ± 1.58	22 ± 0.82
Collagen 10 µg/cm²	15.56 ± 0.4	11.07 ± 0.34	43.06 ± 1.59	20.3 ± 0.82
Collagen 8 µg/cm²	14.72 ± 0.46	9.95 ± 0.32	38.17 ± 1.62	18.7 ± 0.86
Collagen 6 µg/cm²	14.97 ± 0.44	10.74 ± 0.31	37.57 ± 1.53	20.25 ± 0.74
Vitronectin 0.1µg/cm²	14.54 ± 0.56	10.19 ± 0.06	39.41 ± 2.13	22.41 ± 1.29

Attachment

Cell sizes of MG-63 osteoblasts in dependence of adhesion ligands after 24 h cultivation time (Figure 48)

Treatment	Cell size [µm]			
	Nucleus		Cell	
	Length (L_n)	Width (W_n)	Length (L_c)	Width (W_c)
Control	21.37 ± 0.4	14.66 ± 0.31	111.12 ± 4.12	26.71 ± 1.3
Laminin 2 µg/cm²	21.58 ± 0.31	14.89 ± 0.26	115.99 ± 3.84	23.22 ± 0.76
Laminin 1 µg/cm²	21.29 ± 0.63	14.61 ± 0.48	94.83 ± 4.94	29.63 ± 1.42
Fibronectin 5 µg/cm²	21.67 ± 0.51	16.05 ± 0.44	105.78 ± 4.33	25.59 ± 0.85
Fibronectin 3 µg/cm²	19.06 ± 0.61	13.27 ± 0.46	95.84 ± 7.15	26.38 ± 1.37
Fibronectin 1 µg/cm²	20.26 ± 0.67	13.36 ± 0.56	94.15 ± 6.53	25.87 ± 2.02
Collagen 10 µg/cm²	21.6 ± 0.38	15.64 ± 0.3	107.77 ± 3.57	25.24 ± 0.8
Collagen 8 µg/cm²	19.81 ± 0.53	13.19 ± 0.29	93.05 ± 5.18	26.25 ± 1.41
Collagen 6 µg/cm²	20.35 ± 0.54	13.62 ± 0.42	90.76 ± 5.24	27.04 ± 1.15
Vitronectin 0.1µg/cm²	21.72 ± 0.4	15.37 ± 0.32	115.11 ± 3.49	24.81 ± 0.97

Cell sizes of A10 smooth muscle cells in dependence of adhesion ligands after 24 h cultivation time (Figure 49)

Treatment	Cell size [µm]			
	Nucleus		Cell	
	Length (L_n)	Width (W_n)	Length (L_c)	Width (W_c)
Control	26.64 ± 0.62	16.23 ± 0.32	128.26 ± 4.33	63.21 ± 1.92
Laminin 2 µg/cm²	23.14 ± 0.59	16.46 ± 0.29	122.67 ± 4.37	59.57 ± 1.95
Laminin 1 µg/cm²	25.74 ± 0.53	16.25 ± 0.4	90.51 ± 3.29	51.27 ± 2.09
Fibronectin 5 µg/cm²	24.23 ± 0.53	16.82 ± 0.29	137.55 ± 4.81	63.15 ± 2.07
Fibronectin 3 µg/cm²	25.35 ± 0.43	16.25 ± 0.42	82.36 ± 2.64	48.09 ± 1.85
Fibronectin 1 µg/cm²	24.21 ± 0.57	15.98 ± 0.43	76.91 ± 3.02	45.98 ± 1.94
Collagen 10 µg/cm²	24.76 ± 0.54	17.29 ± 0.25	143.61 ± 4.56	61.14 ± 2.07
Collagen 8 µg/cm²	26.4 ± 0.72	17.17 ± 0.52	89.87 ± 3.47	48.34 ± 2.2
Collagen 6 µg/cm²	26.6 ± 0.57	18.5 ± 0.48	90,21 ± 2.96	54.83 ± 2.07
Vitronectin 0.1µg/cm²	26.53 ± 0.51	17.49 ± 0.31	147.65 ± 3.71	58.29 ± 1.67

7.8 Acknowledgements

I would like to thank Prof. Dr. B. Chichkov (Laser Zentrum Hannover e. V., Germany) and Prof. Dr. A. Ngezahayo (Institute of Biophysics, Leibniz University Hannover, Germany) for supervising my dissertation. For providing the materials and laser-fabrication I also thank Dr. A. Ovsianikov, Dipl.-Phys. E. Fadeeva, Dipl.-Ing. M. Grüne, Dr. L. Koch (Laser Zentrum Hannover e. V., Germany) and Dipl.-Chem. S. Harling (Institute of Technical Chemistry, TU Braunschweig, Germany). This work was performed during a scholarship at the European Graduate College „Interference and Quantum Applications". Thanks a lot for the time at the Institute of Biophysics.

Finally, I would like to express my gratitude to my family and boyfriend who always supported me.

Die VDM Verlagsservicegesellschaft sucht für wissenschaftliche Verlage abgeschlossene und herausragende

Dissertationen, Habilitationen, Diplomarbeiten, Master Theses, Magisterarbeiten usw.

für die kostenlose Publikation als Fachbuch.

Sie verfügen über eine Arbeit, die hohen inhaltlichen und formalen Ansprüchen genügt, und haben Interesse an einer honorarvergüteten Publikation?

Dann senden Sie bitte erste Informationen über sich und Ihre Arbeit per Email an *info@vdm-vsg.de*.

Sie erhalten kurzfristig unser Feedback!

VDM Verlagsservicegesellschaft mbH
Dudweiler Landstr. 99   Telefon +49 681 3720 174
D - 66123 Saarbrücken   Fax   +49 681 3720 1749
www.vdm-vsg.de

Die VDM Verlagsservicegesellschaft mbH vertritt

Printed by Books on Demand GmbH, Norderstedt / Germany